Hans F. Weinberger
University of Minnesota
Minneapolis, Minnesota

Variational Methods for Eigenvalue Approximation

siam.
SOCIETY FOR INDUSTRIAL AND APPLIED MATHEMATICS

PHILADELPHIA

10 9 8 7 6 5 4 3

ISBN 0-89871-012-X

VARIATIONAL METHODS FOR EIGENVALUE APPROXIMATION

Contents

Foreword v

Chapter 1
PROLOGUE: WHY STUDY EIGENVALUES?
1. SEPARATION OF VARIABLES 1
2. EIGENVALUES AND NONLINEAR PROBLEMS 6
3. SOME OTHER LINEARIZED PROBLEMS OF VIBRATION AND STABILITY 15

Chapter 2
THE SETTING: LINEAR VECTOR SPACES AND THEIR PROPERTIES
1. LINEAR VECTOR SPACES 19
2. LINEAR TRANSFORMATIONS 20
3. SESQUILINEAR AND QUADRATIC FUNCTIONALS 22
4. HILBERT SPACE 23
5. THE SOBOLEV SPACES 28

Chapter 3
THE EXISTENCE AND CHARACTERIZATION OF EIGENVALUES
1. FORMULATION OF THE SELF-ADJOINT EIGENVALUE PROBLEM 31
2. THE EIGENVALUES AS SUCCESSIVE MAXIMA 38
3. COMPLETE CONTINUITY AND FOURIER SERIES 50
4. INHOMOGENEOUS EQUATIONS 53
5. THE POINCARÉ AND COURANT–WEYL PRINCIPLES .. 55
6. A MAPPING PRINCIPLE 57
7. THE FIRST MONOTONICITY PRINCIPLE AND THE RAYLEIGH–RITZ METHOD 58
8. THE SECOND MONOTONICITY PRINCIPLE 62
9. A SEPARATION THEOREM AND AN INEQUALITY FOR THE EIGENVALUES OF A SUM OF OPERATORS 63

Chapter 4
IMPROVABLE BOUNDS FOR EIGENVALUES
1. A PRIORI AND A POSTERIORI ERROR BOUNDS 67
2. ERROR BOUNDS FOR EIGENVALUES IN THE RAYLEIGH–RITZ METHOD 67

3. THE WEINSTEIN–ARONSZAJN THEOREMS ON FINITE-DIMENSIONAL PERTURBATION 74
4. THE METHOD OF A. WEINSTEIN 81
5. THE METHOD OF N. ARONSZAJN 84
6. BAZLEY'S CHOICE 88
7. TRUNCATION 89
8. ERROR ESTIMATES FOR THE WEINSTEIN–ARONSZAJN METHODS 91
9. COMPLEMENTARY BOUNDS IN TERMS OF MATRIX EIGENVALUES 98
10. SIMPLER UPPER AND LOWER BOUNDS 109
11. INCLUSION THEOREMS 113
12. THE METHOD OF G. FICHERA 115

Chapter 5
EIGENVECTOR APPROXIMATION
1. ERROR BOUNDS FOR EIGENVECTORS 123

Chapter 6
FINITE DIFFERENCE EQUATIONS
1. FINITE DIFFERENCE EQUATIONS FROM THE RAYLEIGH–RITZ METHOD 127
2. THE METHOD OF FINITE ELEMENTS 129
3. UPPER AND LOWER BOUNDS FROM FINITE DIFFERENCE EQUATIONS 133

Chapter 7
SOME OTHER BOUNDS
1. SOME BOUNDS FOR MEMBRANE PROBLEMS 141
2. SYMMETRIZATION 145
3. NON-SELF-ADJOINT PROBLEMS 149

References 151

Index 157

Foreword

This book is based on a series of lectures presented at the N.S.F.–C.B.M.S. Regional Conference on Approximation of Eigenvalues of Differential Operators held during June 26–30, 1972 at Vanderbilt University, Nashville, Tennessee.

The invitation to present this lecture series gave me the opportunity to rethink, rewrite, and complete a large body of material, much of which was previously written only in the form of a technical note and lecture notes. Among the material I am thus able to make available to a wider audience are the complementary bounds from matrix eigenvalues which are presented in §§ 9 and 10 of Chapter 4.

In preparing this manuscript I have attempted to provide a common setting for various methods of bounding the eigenvalues of a self-adjoint linear operator and to emphasize their relationships. The mapping principle of Chapter 3, § 6 serves as a link to connect many of these methods.

Some of the material in this book is new. This is true of the error bounds of §§ 2 and 8 of Chapter 4. A new proof of Aronszajn's rule (Theorem 3.3 of Chapter 4) is also presented.

I wish to thank the National Science Foundation for making the Regional Conference at which these lectures were presented possible, for providing an incentive to overcome my natural laziness, and for sponsoring the research which led to several of the results contained herein. My thanks also go to Philip Crooke and Robert Miura for their smooth and efficient organization of the Conference, and to the typists, Carolyn Lott and Claudia Cuzzort. I am particularly grateful to Gaylord Schwartz who wrote up the 1962 lecture notes on which this manuscript is based and to Robert Miura and Philip Crooke who edited the manuscript.

It is, of course, not possible to present all results on eigenvalue approximation in a book of this size. To those authors whose work I have not included I can only offer my apology and the cold comfort of knowing that I have also skipped some of my own papers on the subject.

HANS F. WEINBERGER

CHAPTER 1

Prologue: Why Study Eigenvalues?

1. Separation of variables. The study of eigenvalue problems has its roots in the method of separation of variables.

We consider a linear class C of functions $u(t_1, \cdots, t_M, x_1, \cdots, x_N)$ of two sets of variables with the following properties:

(a) A function in C is defined whenever $(t_1, \cdots, t_M)$ lies in a set T and $(x_1, \cdots, x_N)$ lies in a set X.

(b) There is a linear class C_T of functions $v(t_1, \cdots, t_M)$ defined on T and a linear class C_X of functions $w(x_1, \cdots, x_N)$ defined on X so that if v is in C_T and w is in C_X, then vw is in C.

(c) Every function u of C is the limit (in some sense) of linear combinations of products $v_i w_i$ with v_i in C_T and w_i in C_X. (C is said to be the tensor product of C_T and C_X.)

We are given linear operators M and N on C which factor in the following sense: There are operators R and S defined on C_T and operators A and B defined on C_X so that for v in C_T and w in C_X

$$M[vw] = R[v]B[w],$$

$$N[vw] = S[v]A[w].$$

We shall write

$$M = RB, \qquad N = SA.$$

We seek particular solutions of the equation

$$RBu + SAu = 0. \tag{1.1}$$

We observe that if we can find a constant λ and an element w in C_X for which

$$Aw = \lambda Bw, \tag{1.2}$$

then the product $u = vw$ is a solution of (1.1) if and only if

$$Rv + \lambda Sv = 0. \tag{1.3}$$

The problem (1.1) on C thus splits into two problems on smaller subspaces. A value λ for which (1.2) has a solution $w \neq 0$ is called an eigenvalue, and w is called the corresponding eigenfunction.

Of course, in order to use the above ideas one must try to show that any solution of (1.1) can be written as a limit of linear combinations of such product solutions.

We shall give some simple examples.

Example 1.1. Let A and B be $L \times L$ matrices. We seek to determine a vector-valued solution $\mathbf{u}$ of the initial value problem for the system of ordinary differential equations

$$(1.4)\qquad \begin{aligned} B\,d\mathbf{u}/dt + A\mathbf{u} &= 0, \\ \mathbf{u}(0) &= \mathbf{u}_0. \end{aligned}$$

We can think of $\mathbf{u}(t)$ written in terms of its components $u_i(t)$ or $u(t, i)$. Thus, u is a function of the continuous variable t and the discrete variable i. The operator d/dt acts only on the t-variable while A and B act on the i-variable. A product solution has the form $u(t, i) = v(t)w(i)$ or, in vector notation,

$$\mathbf{u}(t) = v(t)\mathbf{w},$$

where $v(t)$ is a scalar function and $\mathbf{w}$ is a vector which is independent of t. The equation (1.2) becomes

$$(1.5)\qquad A\mathbf{w} = \lambda B\mathbf{w},$$

while (1.3) gives

$$dv/dt + \lambda v = 0,$$

whose general solution is

$$v = v_0\, e^{-\lambda t}.$$

Calling the eigenvalues of the problem (1.5) $\lambda_1, \lambda_2, \cdots$ and the corresponding eigenvectors $\mathbf{w}_1, \mathbf{w}_2, \cdots$, we see that we have product solutions $e^{-\lambda_n t}\mathbf{w}_n$. We seek to solve the initial value problem (1.4) by a linear combination

$$\mathbf{u}(t) = \sum c_n\, e^{-\lambda_n t}\mathbf{w}_n.$$

The differential equation is satisfied, and we need to choose the coefficients c_n to satisfy the initial condition

$$\sum c_n \mathbf{w}_n = \mathbf{u}_0,$$

where $\mathbf{u}_0$ is given. This equation can be solved for arbitrary data $\mathbf{u}_0$ if and only if the number of linearly independent eigenvectors is equal to the dimension L of the vector space.

We note that λ is an eigenvalue if and only if

$$\det(A - \lambda B) = 0.$$

If the matrix B is nonsingular, this is an equation of degree L. If its roots are distinct, it is easy to see that the L corresponding eigenvectors are linearly independent. If some λ is a root of multiplicity m, the number of corresponding linearly independent eigenvectors is less than or equal to m.

If there are L linearly independent eigenvectors $\mathbf{w}_n$, we form the matrix W whose columns are these eigenvectors. Then all the eigenvalue equations (1.5) together are equivalent to the matrix equation

$$AW = BW\Lambda,$$

where Λ is the diagonal matrix with entries $\lambda_1, \lambda_2, \cdots, \lambda_L$. If we introduce the new variables

$$\tilde{\mathbf{u}}(t) = W^{-1}\mathbf{u}(t),$$

the differential equation in (1.4) becomes

$$BW\, d\tilde{\mathbf{u}}/dt + AW\tilde{\mathbf{u}} = 0$$

or

$$BW[d\tilde{\mathbf{u}}/dt + \Lambda\tilde{\mathbf{u}}] = 0.$$

If B is nonsingular, we may take linear combinations by operating with $(BW)^{-1}$. We thus arrive at the equivalent system

$$d\tilde{\mathbf{u}}/dt + \Lambda\tilde{\mathbf{u}} = 0.$$

Since Λ is diagonal, the nth equation involves only $\tilde{u}_n$, so that the system is uncoupled. This makes it very easy to solve. In fact, we can also operate in the same way on the inhomogeneous system

$$B\, d\mathbf{u}/dt + A\mathbf{u} = \mathbf{f}.$$

It thus reduces to the uncoupled system

$$d\tilde{\mathbf{u}}/dt + \Lambda\tilde{\mathbf{u}} = (BW)^{-1}\mathbf{f}.$$

We have used here the fact that the matrix transformation $M \to (BW)^{-1}MW$ simultaneously reduces the matrices A and B to diagonal form. The L linearly independent eigenvectors lead to this simultaneous diagonalization. If they do not exist but B is nonsingular, then there are solutions of the differential equation (1.4) of the form $\sum t^j e^{-\lambda t}\mathbf{w}^{(j)}$ which are not sums of product solutions.

Example 1.2. Consider the problem of heat flow in an insulated container

$$\begin{aligned} &\partial u/\partial t - \Delta u = 0 \quad \text{in } D,\\ &\partial u/\partial n = 0 \qquad\quad\ \text{on } \partial D, \qquad\qquad (1.6)\\ &u(0, x_1, x_2, x_3) = u_0(x_1, x_2, x_3) \quad \text{in } D.\end{aligned}$$

Here D is a bounded three-dimensional domain with boundary ∂D, and $u_0(x_1, x_2, x_3)$ is prescribed. The directional derivative along the outward normal to the boundary is represented by $\partial u/\partial n$ and Δ is the usual Laplace operator. The class C_X is taken to be functions $w(x_1, x_2, x_3)$ which satisfy the homogeneous condition $\partial w/\partial n = 0$ on ∂D. We seek product solutions $u = v(t)w(x_1, x_2, x_3)$ of the differential equation in (1.6). Equation (1.2) becomes

$$\begin{aligned} -\Delta w &= \lambda w \quad \text{in } D,\\ \partial w/\partial n &= 0 \quad\ \ \text{on } \partial D, \end{aligned} \qquad (1.7)$$

and (1.3) becomes

$$dv/dt + \lambda v = 0.$$

Thus if λ_n is an eigenvalue of the problem (1.7) with the corresponding eigenfunction w_n, then $e^{-\lambda_n t}w_n$ is a particular solution of the heat equation. If we wish to find a solution

$$u = \sum c_n e^{-\lambda_n t} w_n(x_1, x_2, x_3) \tag{1.8}$$

of the problem (1.6), we must determine the coefficients c_n so that

$$\sum c_n w_n(x_1, x_2, x_3) = u_0(x_1, x_2, x_3).$$

(There are now infinitely many eigenvalues, so the sums are infinite series.)

If this problem can be solved for arbitrary u_0 in some class, we can formally solve the inhomogeneous problem

$$\begin{aligned} \partial u/\partial t - \Delta u &= f(t, x_1, x_2, x_3), \\ u(0, x_1, x_2, x_3) &= u_0(x_1, x_2, x_3), \end{aligned} \tag{1.9}$$

in the following manner. We seek a solution of the form

$$u(t, x) = \sum \tilde{u}_n(t) w_n(x_1, x_2, x_3),$$

and we suppose that for each fixed t,

$$f(t, x_1, x_2, x_3) = \sum \phi_n(t) w_n(x_1, x_2, x_3).$$

Differentiating term by term and using the fact that w_n is an eigenfunction, we have

$$\sum [d\tilde{u}_n/dt + \lambda_n \tilde{u}_n - \phi_n] w_n = 0.$$

Thus, we have a solution if $\tilde{u}_n$ satisfies

$$\begin{aligned} d\tilde{u}_n/dt + \lambda_n \tilde{u}_n &= \phi_n(t), \\ \tilde{u}_n(0) &= c_n. \end{aligned}$$

Thus, just as in the case of ordinary differential equations, the eigenfunctions w_n allow us to make a change of variables which uncouples the system in the sense that it reduces to ordinary differential equations.

It is easily seen that the eigenvalues of (1.7) are real and nonnegative. The lowest eigenvalue λ_1 is zero with $w_1 \equiv 1$. The others are strictly positive. We then see from (1.8) that as $t \to \infty$, the solution converges exponentially to the constant c_1, which is determined by the initial values. The rate of convergence is determined by the smallest positive eigenvalue λ_2.

Example 1.3. The study of sound wave propagation leads to the problem

$$\begin{aligned} &\partial^2 u/\partial t^2 - \Delta u = 0 \quad \text{in } D, \\ &\partial u/\partial n = 0 \qquad\qquad \text{on } \partial D, \\ &u(0, x_1, x_2, x_3) = u_0(x_1, x_2, x_3) \quad \text{in } D, \\ &\partial u/\partial t(0, x_1, x_2, x_3) = 0. \end{aligned} \tag{1.10}$$

We find product solutions of (1.10) in the form $\exp(\pm i\sqrt{\lambda_n}t) w_n(x_1, x_2, x_3)$ where λ_n and w_n are eigenvalues and eigenfunctions of (1.7). Each of these solutions has fixed "shape" and varies sinusoidally in time, and is called a normal mode. The frequency of oscillation is $\sqrt{\lambda_n}/2\pi$ which is determined by the eigenvalue λ_n.

The inhomogeneous problem

$$\partial^2 u/\partial t^2 - \Delta u = f$$

can be solved by letting

$$u(t, x_1, x_2, x_3) = \sum \tilde{u}_n(t) w_n(x_1, x_2, x_3),$$

$$f(t, x_1, x_2, x_3) = \sum \phi_n(t) w_n(x_1, x_2, x_3).$$

Then

$$d^2\tilde{u}_n/dt^2 + \lambda_n \tilde{u}_n = \phi_n(t).$$

If $\phi_n(t) = a \sin \omega t$, we obtain the particular solution

$$u_n = \frac{1}{\lambda_n - \omega^2} a \sin \omega t.$$

If ω is near $\sqrt{\lambda_n}$, this term may be very large. If $\omega = \sqrt{\lambda_n}$, the particular solution becomes

$$at \sin \omega t,$$

which becomes unbounded in time. This phenomenon is called resonance and the eigenvalues λ_n determine the resonant frequencies.

Example 1.4. We consider Laplace's equation

$$-\Delta u = 0 \tag{1.11}$$

on the product domain

$$D = \{x_1, \cdots, x_N | (x_1, \cdots, x_M) \in \mathscr{R}, (x_{M+1}, \cdots, x_N) \in \mathscr{S}\}$$

with the boundary conditions

$$\begin{aligned} u &= \psi(x_1, \cdots, x_N) \quad \text{for} \quad (x_1, \cdots, x_M) \in \partial\mathscr{R}, \\ u &= 0 \quad \text{for} \quad (x_{M+1}, \cdots, x_N) \in \partial\mathscr{S}. \end{aligned} \tag{1.12}$$

We seek product solutions

$$u(x_1, \cdots, x_N) = v(x_1, \cdots, x_M) w(x_{M+1}, \cdots, x_N),$$

and take for C_X functions defined in $\mathscr{S}$ which vanish on its boundary. Then (1.2) becomes

$$\begin{aligned} -\Delta w &= \lambda w(x_{M+1}, \cdots, x_N) \quad \text{in } \mathscr{S}, \\ w &= 0 \quad \text{on } \partial\mathscr{S}, \end{aligned}$$

while (1.3) gives

$$-\Delta v + \lambda v(x_1, \cdots, x_M) = 0 \quad \text{in } \mathscr{R}.$$

We must then seek to find a series

$$\sum c_n v_n w_n$$

which satisfies the inhomogeneous boundary condition in (1.12).

Example 1.5. The Schroedinger equation for a system of bodies is

$$-i\frac{\partial u}{\partial t} - \Delta u + Vu = 0, \tag{1.13}$$

where u is a complex-valued function and $V(\mathbf{x})$ is a given real-valued function. The number of x-variables is three times the number of bodies. If

$$-\Delta w + Vw = \lambda w \tag{1.14}$$

with λ real, we obtain the product solution $e^{-i\lambda t}w$. The absolute value of this solution is time independent. It represents a stationary state. The eigenvalue λ is its energy.

The problem is frequently to be solved on an unbounded domain or even all of space. We require that the x-integral of $|u|^2$ remain finite. The problem (1.14) may or may not have eigenvalues.

2. Eigenvalues and nonlinear problems. The eigenvalue problems which we shall study are all linear. Nature, on the other hand, tends to produce nonlinear problems. Most linear problems are obtained by neglecting nonlinear effects. In this section we shall show that linearization can give important information about at least some nonlinear problems.

We begin with a straight rigid rod which is attached to a hinge with a torsion spring at its lower end (Fig. 2.1). If the angle between the rod and the vertical direction is θ, the spring exerts a torque $k\theta$ tending to make the rod vertical again. A downward force F acts at the top of the rod. We denote the length of the rod by l and its mass, which we assume to be uniformly distributed, by m.

FIG. 2.1.

We consider the problem of finding the motion of the rod when it is released at time zero with given initial values of θ and $\dot\theta = d\theta/dt$. If we neglect the gravitational force, the equation of motion is easily found to be

$$\tfrac{1}{3}ml^2\ddot\theta + k\theta - lF\sin\theta = 0. \tag{2.1}$$

Multiplying by $\dot\theta$ and integrating, we find the law of conservation of energy

$$T + V = \text{const.}, \tag{2.2}$$

where

$$(2.3) \qquad \begin{aligned} T &= \tfrac{1}{6}ml^2\dot{\theta}^2, \\ V &= \tfrac{1}{2}k\theta^2 - lF(1 - \cos\theta). \end{aligned}$$

It is clear from (2.1) that $\theta \equiv 0$ is a solution. That is, the vertical position represents a position of equilibrium. We wish to consider a small motion about this equilibrium position. Let us take the initial conditions

$$(2.4) \qquad \theta(0) = \varepsilon\alpha, \qquad \theta(0) = \varepsilon\beta,$$

and let us assume that the solution $\theta(t, \varepsilon)$ is a continuously differentiable function of ε at $\varepsilon = 0$. We differentiate (2.1) with respect to ε and set $\varepsilon = 0$. Defining

$$\phi(t) \equiv \frac{\partial\theta}{\partial\varepsilon}(t, 0),$$

we obtain the equation

$$\begin{gathered} \tfrac{1}{3}ml^2\ddot{\phi} + (k - lF)\phi = 0, \\ \phi(0) = \alpha, \qquad \dot{\phi}(0) = \beta. \end{gathered}$$

If we assume that

$$(2.5) \qquad k > lF,$$

we find the solution

$$\phi(t) = \alpha\cos\omega t + \frac{\beta}{\omega}\sin\omega t,$$

where

$$\omega = \left\{\frac{k - lF}{\tfrac{1}{3}ml^2}\right\}^{1/2}.$$

Thus we can expect that when ε is small we have periodic motions with period $2\pi/\omega$. On the other hand, if $k < lF$, we obtain unbounded solutions and we can expect the equilibrium position $\theta = 0$ to be unstable.

In fact, if we substitute the boundary condition (2.4) in the energy integral (2.2), we see that

$$(2.6) \qquad \tfrac{1}{6}ml^2\dot{\theta}^2 = \tfrac{1}{6}ml^2\varepsilon^2\beta^2 + V(\varepsilon\alpha) - V(\theta).$$

We observe that

$$\begin{aligned} V(0) &= V'(0) = 0, \\ V''(\theta) &= k - lF\cos\theta \geqq k - lF. \end{aligned}$$

In this way we see that if $k - lF \geqq 0$, V is convex and positive for $\theta \neq 0$. Therefore, for sufficiently small ε the facts that $\dot{\theta}^2$ is nonnegative and that $V + T$ is constant imply that θ remains bounded: $|\theta| \leqq \bar{\theta}$ where $V(\bar{\theta}) = \tfrac{1}{6}ml^2\varepsilon^2\beta^2 + V(\varepsilon\alpha)$. It is then easily seen that the solution θ is a periodic function with period

$$4(\tfrac{1}{6}ml^2)^{1/2}\int_0^{\tilde{\theta}}(\tfrac{1}{6}ml^2\varepsilon^2\beta^2 + V(\varepsilon\alpha) - V(\theta))^{-1/2}\,d\theta.$$

As $\varepsilon \to 0$ this period approaches $2\pi/\omega$, and $\theta(t, \varepsilon)/\varepsilon$ approaches $\phi(t)$. The equilibrium position $\theta = 0$ is stable in the sense that given any $\delta > 0$ we can make $\theta^2 + \dot{\theta}^2 < \delta$ for all time by making $\theta^2(0) + \dot{\theta}^2(0)$ sufficiently small. Since we have not included any damping in our formulation, the motion does not approach the equilibrium state. That is, this position $\theta = 0$ is not asymptotically stable.

If, on the other hand, $k - lF < 0$, then $V''(0) < 0$. Therefore when ε is small, the set of θ where the right-hand side of (2.6) is positive extends approximately to the point $\tilde{\theta}$ which is the positive solution of the equation $V(\theta) = 0$. It is easily seen that the motion goes near that point so that, as the linear theory indicated, the equilibrium position $\theta = 0$ is unstable when $k < lF$.

Thus, the linear theory has proved useful both in predicting small vibrations and in predicting whether or not a particular equilibrium position is stable.

We can now go further and ask whether there are any equilibrium positions other than $\theta = 0$. We see from (1.1) that such an equilibrium position, say $\theta = \theta^*$, would be characterized by the vanishing of the torque

$$k\theta^* - lF\sin\theta^* = 0$$

or

$$\frac{\sin\theta^*}{\theta^*} = \frac{k}{lF}.$$

Since the function $\sin\theta/\theta$ decreases from 1 to 0 as θ goes from 0 to π we conclude that if $k \geqq lF$ (the stable case), there is no equilibrium position other than $\theta = 0$. On the other hand for $k < lF$ there is exactly one equilibrium position θ^* between 0 and π. Then $-\theta^*$ is another equilibrium position. (For lF/k sufficiently large, there are other equilibrium positions with $|\theta| > \pi$, but we shall not consider them here.)

Thus we see that when the parameter $k - lF$ decreases through the value 0, the unique equilibrium solution $\theta = 0$ bifurcates into the three solutions $\theta = 0$, $\theta = \theta^*$, $\theta = -\theta^*$. The linearized problem predicts this bifurcation by giving an infinite period ($\omega = 0$). This can be interpreted to mean that when $k - lF = 0$, there is an infinitesimally close state to which the solution moves, never to return. It is as close as a linear theory can come to predicting bifurcation, and it does so successfully in many problems.

We now consider the slightly more complicated situation in which a second rod of the same length l and mass m is attached to the first rod by means of a hinge with a torsion spring with the same spring constant k (see Fig. 2.2). The spring tends to keep the two rods aligned. The first rod is attached to the ground by a torsion spring as before, and the vertical force F is now applied at the top of the second rod.

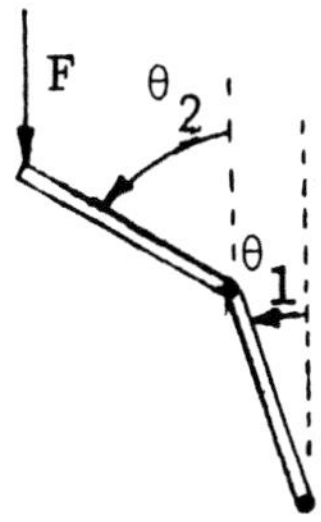

FIG. 2.2.

We compute the kinetic energy T and the potential energy V of this system in terms of the angles θ_1 and θ_2 which the two rods make with the vertical directions. (For the sake of simplicity, we again neglect the gravitational force.)

(2.7)
$$T = \tfrac{1}{6}ml^2[4\dot{\theta}_1^2 + 3\dot{\theta}_1\dot{\theta}_2 \cos(\theta_1 - \theta_2) + \dot{\theta}_2^2],$$
$$V = \tfrac{1}{2}k\theta_1^2 + \tfrac{1}{2}k(\theta_2 - \theta_1)^2 - Fl(2 - \cos\theta_1 - \cos\theta_2).$$

Lagrange's equations are

(2.8)
$$\frac{d}{dt}\left(\frac{\partial T}{\partial \dot{\theta}_i}\right) - \frac{\partial T}{\partial \theta_i} + \frac{\partial V}{\partial \theta_i} = 0$$

with $i = 1, 2$, or

$$\tfrac{1}{6}ml^2[8\ddot{\theta}_1 + 3\ddot{\theta}_2 \cos(\theta_1 - \theta_2) + 3\dot{\theta}_2^2 \sin(\theta_1 - \theta_2)] + k(2\theta_1 - \theta_2) - lF\sin\theta_1 = 0,$$
$$\tfrac{1}{6}ml^2[3\ddot{\theta}_1 \cos(\theta_1 - \theta_2) + 2\ddot{\theta}_2 - 3\dot{\theta}_1^2 \sin(\theta_1 - \theta_2)] + k(\theta_2 - \theta_1) - lF\sin\theta_2 = 0.$$

Clearly, this system again has the equilibrium position $\theta_1 = 0, \theta_2 = 0$.

We now examine the solution which corresponds to the initial values

$$\theta_i(0) = \varepsilon\alpha_i, \qquad \dot{\theta}_i(0) = \varepsilon\beta_i, \qquad i = 1, 2.$$

Again we assume that the solution $\theta_i(t, \varepsilon)$ is differentiable at $\varepsilon = 0$ and let

$$\phi_i(t) \equiv \frac{\partial \theta_i}{\partial \varepsilon}(t, 0), \qquad i = 1, 2.$$

Differentiating Lagrange's equations (2.8) with respect to ε and setting $\varepsilon = 0$, we find the linearized system

(2.9)
$$\sum_{j=1}^{2}\left\{\left[\frac{\partial^2 T}{\partial \dot{\theta}_i \partial \dot{\theta}_j}(0)\ddot{\phi}_j\right] + \frac{\partial^2 V}{\partial \theta_i \partial \theta_j}(0)\phi_j\right\} = 0, \qquad i = 1, 2.$$

The argument 0 in the derivatives of T and V means that we are to set $\theta_j = \dot{\theta}_j = 0$. Since T is quadratic in $\dot{\theta}$, all second derivatives of T which involve the variables θ_j vanish when we put $\dot{\theta} = 0$.

We define the coefficient matrices

$$A_{ij} = \frac{\partial^2 V}{\partial\theta_i \partial\theta_j}(0), \qquad B_{ij} = \frac{\partial^2 T}{\partial\dot{\theta}_i \partial\dot{\theta}_j}(0). \tag{2.10}$$

(These coefficients are easily calculated from the definition (2.7).) Then the system (2.9) can be written in vector form as

$$B\ddot{\boldsymbol{\phi}} + A\boldsymbol{\phi} = 0. \tag{2.11}$$

We observe that the matrices A and B are symmetric. Moreover, if $\boldsymbol{\xi}$ is any vector, the quadratic form $\boldsymbol{\xi} \cdot B\boldsymbol{\xi}$ approximates ε^{-2} times the kinetic energy T corresponding to the state $\boldsymbol{\theta} = \mathbf{0}, \dot{\boldsymbol{\theta}} = \varepsilon\boldsymbol{\xi}$. Therefore, $\boldsymbol{\xi} \cdot B\boldsymbol{\xi} > 0$ whenever $\boldsymbol{\xi} \neq \mathbf{0}$. We say that B is *positive definite*.

The problem (2.11) can be solved by separation of variables. Putting $\boldsymbol{\phi} = v(t)\mathbf{w}$, we have

$$A\mathbf{w} = \lambda B\mathbf{w}, \tag{2.12}$$

$$\ddot{v} + \lambda v = 0. \tag{2.13}$$

The eigenvalue equation (2.12) has a nontrivial solution w if and only if

$$\det(A - \lambda B) = 0.$$

Taking account of (2.7) and (2.10), we find that this equation is

$$\begin{vmatrix} 2k - lF - \frac{4}{3}ml^2\lambda & -k - \frac{1}{2}ml^2\lambda \\ -k - \frac{1}{2}ml^2\lambda & k - lF - \frac{1}{3}ml^2\lambda \end{vmatrix} = 0.$$

The discriminant of this equation is nonnegative, so that the roots are always real. Moreover, if $F > 0$ and $k > 0$ the discriminant is positive. Hence there are two distinct roots $\lambda_1 < \lambda_2$.

If the eigenvalues are both positive, the problem (2.11) has solutions which are sums of normal modes of the form $\exp(\pm i\sqrt{\lambda_n}t)\mathbf{w}_n$. The periods are $2\pi/\sqrt{\lambda_n}$. The linearized theory predicts that if both eigenvalues are positive, the position $\boldsymbol{\theta} = 0$ is stable and that the motion corresponding to small initial displacements is almost periodic. If one of the eigenvalues is negative, the linear theory predicts instability. We wish to verify these predictions.

Suppose first that both eigenvalues are positive. Let Λ be the 2×2 diagonal matrix whose entries are the eigenvalues λ_1 and λ_2 of the problem $A\mathbf{w} = \lambda B\mathbf{w}$. Let W be the matrix whose two columns are eigenvectors $\mathbf{w}$ corresponding to these eigenvalues. Then

$$AW = BW\Lambda.$$

Premultiplying by the transpose W^* of W, we have

$$W^*AW = W^*BW\Lambda.$$

Taking transposes of both sides and recalling that A and B are symmetric, we see that

$$\Lambda W^*BW = W^*AW = W^*BW\Lambda.$$

That is, Λ commutes with the symmetric matrix W^*BW. Writing this fact in terms of components, we have

$$\lambda_i(W^*BW)_{ij} = (W^*BW)_{ij}\lambda_j$$

for $i, j = 1, 2$. Then either $\lambda_1 = \lambda_2$ or the off-diagonal elements $(W^*BW)_{12}$ and $(W^*BW)_{21}$ are zero. In either case, we see that also

$$(W^*BW)_{ij}\lambda_j = \lambda_i^{1/2}(W^*BW)_{ij}\lambda_j^{1/2}.$$

We define $\Lambda^{1/2}$ to be the diagonal matrix whose entries are the positive square roots $\lambda_1^{1/2}$ and $\lambda_2^{1/2}$. We then have

$$W^*AW = \Lambda^{1/2}W^*BW\Lambda^{1/2},$$

or

$$A = (W\Lambda^{1/2}W^{-1})^*B(W\Lambda^{1/2}W^{-1}).$$

Thus for any real nonzero vector $\boldsymbol{\xi}$

$$\begin{aligned}\boldsymbol{\xi}\cdot A\boldsymbol{\xi} &= \boldsymbol{\xi}\cdot(W\Lambda^{1/2}W^{-1})^*B(W\Lambda^{1/2}W^{-1})\boldsymbol{\xi}\\ &= (W\Lambda^{1/2}W^{-1}\boldsymbol{\xi})\cdot B(W\Lambda^{1/2}W^{-1}\boldsymbol{\xi}) > 0\end{aligned}$$

since B is positive definite. That is, A is also positive definite.

Recalling the definition (2.10) of A, we see that for small $\boldsymbol{\theta}$

$$V(\boldsymbol{\theta}) = \tfrac{1}{2}\boldsymbol{\theta}\cdot A\boldsymbol{\theta} + O(|\boldsymbol{\theta}|^3).$$

Thus, we can find positive constants η and δ so that

$$V(\boldsymbol{\theta}) \geqq \eta|\boldsymbol{\theta}|^2 \quad \text{for } |\boldsymbol{\theta}| \leqq \delta.$$

Conservation of energy now implies, as in the case of a single rod, that the position $\boldsymbol{\theta} = 0$ is stable. Moreover, we can show that the solution of (2.8) corresponding to initial conditions $\boldsymbol{\theta}(0) = \varepsilon\boldsymbol{\alpha}, \dot{\boldsymbol{\theta}}(0) = \varepsilon\boldsymbol{\beta}$ is differentiable in ε so that for any finite time interval, $\varepsilon\boldsymbol{\phi}(t)$ is a good approximation to $\boldsymbol{\theta}(t, \varepsilon)$ when ε is sufficiently small. A closer examination of V shows that when both eigenvalues are positive, V is strictly convex, so that $\boldsymbol{\theta} = 0$ is the only equilibrium position.

If one of the eigenvalues, say λ_1, is negative, we see from the eigenvalue equation

$$A\mathbf{w}_1 = \lambda_1 B\mathbf{w}_1$$

that

$$\mathbf{w}_1\cdot A\mathbf{w}_1 = \lambda_1\mathbf{w}_1\cdot B\mathbf{w}_1 < 0.$$

Therefore,

$$V(\varepsilon\mathbf{w}_1) = \tfrac{1}{2}\varepsilon^2\mathbf{w}_1\cdot A\mathbf{w}_1 + O(\varepsilon^3)$$

is negative for ε sufficiently small. It follows that the state $\boldsymbol{\theta} = 0$ is actually unstable, as predicted by the linearized theory.

Since V is strictly convex when the eigenvalues are positive, we see that for any $\delta > 0$, V is positive on the circle $|\boldsymbol{\theta}| = \delta$. By continuity there is a positive constant ν so that the same is still true for $\lambda_1 > -\nu$. Then if $-\nu < \lambda_1 < 0$, V becomes negative near $\boldsymbol{\theta} = 0$ but is positive at $|\boldsymbol{\theta}| = \delta$. It follows that V must attain a negative minimum value at a point $\boldsymbol{\theta} = \boldsymbol{\theta}^*$ of the disc $|\theta| < \delta$. At this point grad $V = 0$, so that $\boldsymbol{\theta} = \boldsymbol{\theta}^*$ is an equilibrium solution of (2.8). By symmetry, $-\boldsymbol{\theta}^*$ is also an equilibrium point. The value of $\boldsymbol{\theta}^*$ clearly depends on λ_1, and since δ is arbitrary, it approaches zero with λ_1. We can say that as λ_1 is decreased (by increasing F or decreasing k) through zero, the unique solution $\boldsymbol{\theta} = \mathbf{0}$ bifurcates into the three solutions $\mathbf{0}$ and $\pm\boldsymbol{\theta}^*$. The latter two are stable.

The bifurcation occurs when $\lambda_1 = 0$. This condition is characterized by the equation det $A = 0$ or

$$\begin{vmatrix} 2k - lF & -k \\ -k & k - lF \end{vmatrix} = 0.$$

If we fix k and vary F, the critical value of F is an eigenvalue of the matrix $\begin{pmatrix} 2k & -k \\ -k & k \end{pmatrix}$ with respect to the matrix lI. The critical value of F is the vertical force which makes the system buckle out of its vertical position. A problem in which one wishes to find the value of a parameter which describes the force needed to produce buckling is called a *buckling problem*.

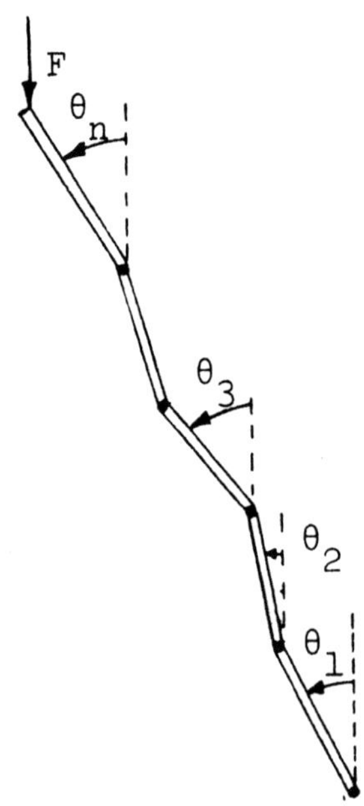

FIG. 2.3.

We can easily extend the above analysis to a more complicated mechanical system. For instance, we could replace the linear torsion springs by nonlinear ones. Instead we consider the problem of n rigid rods of length l and mass m connected as before by hinges with linear torsion springs having spring constant k (see Fig. 2.3). The kinetic energy T and potential energy V are given (if the gravitational force is neglected) by

$$\begin{aligned} T &= \frac{ml^2}{12} \sum_{i,j=1}^{n} [6n + 3 - 6\max(i,j) - \delta_{ij}]\dot{\theta}_i\dot{\theta}_j \cos(\theta_i - \theta_j), \\ V &= \frac{1}{2}k\left\{\theta_1^2 + \sum_{j=2}^{n} (\theta_j - \theta_{j-1})^2\right\} - Fl\left(n - \sum_{j=1}^{n} \cos\theta_j\right). \end{aligned} \tag{2.14}$$

We can again write Lagrange's equations in the form (2.8) with i running from 1 to n. Clearly $V(\mathbf{0}) = \operatorname{grad} V(\mathbf{0}) = 0$, so that $\boldsymbol{\theta} \equiv \mathbf{0}$ is an equilibrium solution. The linearized equation for small perturbations $\boldsymbol{\phi}$ from this equilibrium are again given by the equation (2.11) with the symmetric matrices A and B defined by (2.10). Again we find that the position $\boldsymbol{\theta} = \mathbf{0}$ is stable if and only if all the eigenvalues of the problem

$$\det(A - \lambda B) = 0$$

are positive. (We shall demonstrate in the next chapter that they are real.) Moreover, the solution $\boldsymbol{\theta} = \mathbf{0}$ bifurcates when one of the eigenvalues becomes zero so that $\det A = 0$.

We now pass to the limit as the number n of rods approaches infinity while the total length nl and the total mass nm retain fixed values $\bar{l}$ and $\bar{m}$, respectively, and the ratio k/n approaches the limit $\bar{k}$. The system becomes an inextensible but flexible rod or leaf spring. The unknown becomes the angle $\theta(s, t)$ with the vertical at time t at a point at distance s from the lower end.

We then see from (2.14) that

$$\begin{aligned} T &= \frac{1}{2}\bar{m}\int_0^{\bar{l}}\int_0^{\bar{l}}\left[1 - \frac{\max(s_1, s_2)}{\bar{l}}\right]\dot{\theta}(s_1, t)\dot{\theta}(s_2, t)\cos(\theta(s_1) - \theta(s_2))\, ds_1\, ds_2, \\ V &= \frac{1}{2}\bar{k}\bar{l}\int_0^{\bar{l}}\left(\frac{\partial\theta}{\partial s}\right)^2 ds - F\int_0^{\bar{l}}[1 - \cos\theta]\, ds, \end{aligned} \tag{2.15}$$

where the dot denotes partial differentiation with respect to t.

Lagrange's equations become

$$\begin{aligned} \bar{m}\int_0^{\bar{l}}\left[1 - \frac{\max(s, s_2)}{\bar{l}}\right]\Big[&\ddot{\theta}(s_2, t)\cos(\theta(s, t) - \theta(s_2, t)) \\ &+ \dot{\theta}(s_2, t)^2 \sin(\theta(s, t) - \theta(s_2, t))\Big]\, ds_2 \\ &- \bar{k}\bar{l}\frac{\partial^2\theta}{\partial s^2} - F\sin\theta = 0. \end{aligned} \tag{2.16}$$

This is a rather formidable integro-differential equation. From the original problem we see that $\theta = 0$ at the bottom $s = 0$. Taking the limit of equation (2.8) with $i = n$ divided by l we see that $\partial\theta/\partial s = 0$ at the top $s = l$. Thus, we have the boundary conditions

$$\theta(0, t) = \frac{\partial\theta}{\partial s}(l, t) = 0. \tag{2.17}$$

Clearly the equation (2.16) with the boundary conditions (2.17) has the equilibrium solution $\theta \equiv 0$. We now examine the small perturbation obtained from the initial conditions

$$\theta(s, 0) = \varepsilon\alpha(s), \qquad \dot{\theta}(s, 0) = \varepsilon\beta(s).$$

As before, we assume (and we can prove) that the solution of this problem is continuously differentiable in ε and vanishes at $\varepsilon = 0$. We define

$$\phi(s, t) \equiv \left.\frac{\partial\theta}{\partial\varepsilon}\right|_{\varepsilon=0}.$$

Differentiating the equation (2.16) and setting $\varepsilon = 0$, we obtain the linear integro-differential equation

$$\bar{m}\int_0^l \left[1 - \frac{\max(s, s_2)}{l}\right]\ddot{\phi}(s_2, t)\, ds_2 - \bar{k}l\frac{\partial^2\phi}{\partial s^2} - F\phi = 0. \tag{2.18}$$

We differentiate twice with respect to s to obtain the differential equation

$$-\frac{\bar{m}}{l}\frac{\partial^2\phi}{\partial t^2} - \bar{k}l\frac{\partial^4\phi}{\partial s^4} - F\frac{\partial^2\phi}{\partial s^2} = 0. \tag{2.19}$$

This is the usual linearized equation for a vibrating beam.

In addition to the initial conditions

$$\phi(s, 0) = \alpha(s), \qquad \dot{\phi}(s, 0) = \beta(s), \tag{2.20}$$

we need boundary conditions. Two of these are supplied by the conditions (2.17). Two more come from the fact that the integral in (2.18) vanishes at $s = l$, while the derivative of the integral with respect to s vanishes at $s = 0$. Thus we find that

$$\begin{aligned} \phi = \bar{k}l\frac{\partial^3\phi}{\partial s^3} + F\frac{\partial\phi}{\partial s} &= 0 \quad \text{at } s = 0, \\ \frac{\partial\phi}{\partial s} = \bar{k}l\frac{\partial^2\phi}{\partial s^2} + F\phi &= 0 \quad \text{at } s = l. \end{aligned} \tag{2.21}$$

It is more usual to work with the independent variable

$$x(s, t) = \int_0^s \phi(\sigma, t)\, d\sigma,$$

which represents the horizontal displacement of the point at distance s from the end. This function again satisfies the differential equation (2.19), but the boundary conditions become

$$x = \frac{\partial x}{\partial s} = 0 \quad \text{at } s = 0,$$

$$\frac{\partial^2 x}{\partial s^2} = \bar{k}l\frac{\partial^3 x}{\partial s^3} + F\frac{\partial x}{\partial s} = 0 \quad \text{at } s = l.$$

Separation of variables produces product solutions $\exp(\pm i\sqrt{\lambda_n}t)\psi_n(x)$ of the system (2.19), (2.21), where λ_n and ψ_n are eigenvalues and eigenfunctions of the problem

$$\frac{\bar{k}l^2}{\bar{m}}\frac{d^4\psi}{ds^4} + F\frac{l}{\bar{m}}\frac{d^2\psi}{ds^2} - \lambda\psi = 0,$$

(2.22) $$\psi = \bar{k}l\frac{d^3\psi}{ds^3} + F\frac{d\psi}{ds} = 0 \quad \text{at } s = 0,$$

$$\frac{d\psi}{ds} = \bar{k}l\frac{d^2\psi}{ds^2} + F\psi = 0 \quad \text{at } s = l.$$

When all the eigenvalues of this problem are positive, we expect (and can prove) stability of the solution $\theta = 0$. When one of the eigenvalues becomes negative, we can expect the position $\theta = 0$ to become unstable and to bifurcate into other equilibrium states. To detect the onset of bifurcation, we set $\lambda = 0$ in (2.22). We may then integrate twice, using the end conditions to arrive at the buckling problem

$$kl\frac{d^2\psi}{ds^2} + F\psi = 0,$$

$$\psi(0) = 0, \qquad \psi'(l) = 0.$$

This is an eigenvalue problem for the critical value of F.

Linear analysis alone does not prove stability, instability, uniqueness, or bifurcation for the nonlinear problem. However, these proofs can be carried out for many nonlinear problems by the methods of nonlinear functional analysis (see, e.g., Sattinger (1973)).

3. Some other linearized problems of vibration and stability. There are many other physical problems which, when linearized, lead to eigenvalue problems. We shall mention a few of them here to give an idea of the scope of our methods.

(a) *Vibrations of an elastic solid.* If the equations of elasticity are linearized, one is led to stress–strain relations of the form

(3.1) $$\sigma_{ij} = \frac{1}{2}\sum_{k,l=1}^{3} A_{ijkl}(x_1, x_2, x_3)\left(\frac{\partial u_k}{\partial x_l} + \frac{\partial u_l}{\partial x_k}\right).$$

The tensor A describes the elastic properties of the material at the point (x_1, x_2, x_3). It has the symmetry properties $A_{ijkl} = A_{jikl} = A_{ijlk} = A_{klij}$. The vector $\mathbf{u}(x_1, x_2, x_3, t)$ represents the displacement from a neutral position of the material. If the material has density $\rho(x_1, x_2, x_3)$, the equations of motion are

$$\sum_{j=1}^{3} \frac{\partial \sigma_{ij}}{\partial x_j} = \rho \frac{\partial^2 u_i}{\partial t^2}$$

with σ_{ij} given by (3.1). Separation of variables gives normal modes of the form $\exp(\pm i\sqrt{\lambda}t)\mathbf{w}(x)$, where λ and $\mathbf{w}$ are to be determined from the system of partial differential equations

$$-\frac{1}{2}\sum_{j,k,l=1}^{3} \frac{\partial}{\partial x_j}\left[A_{ijkl}\left(\frac{\partial w_k}{\partial x_l} + \frac{\partial w_l}{\partial x_k}\right)\right] = \lambda \rho w_i, \qquad i = 1, 2, 3, \tag{3.2}$$

with suitable boundary conditions.

The boundary conditions typically equate the displacement $\mathbf{w}$, the force $\sum \sigma_{ij} n_j$, or some combination thereof to zero. (Here $\mathbf{n}$ is the outward unit normal on the boundary.)

(b) *The vibrating plate.* If we deal with a thin elastic plate which is assumed to be homogeneous and isotropic, the equation of motion for the normal component of displacement $u(x_1, x_2, t)$ is approximately

$$\Delta^2 u + \frac{\partial^2 u}{\partial t^2} = 0,$$

where Δ^2 represents the biharmonic operator, which is the Laplace operator applied twice. (As always, we make as many physical constants as possible equal to one by a proper choice of units.) Separation of variables now gives normal modes of the form $\exp(i\sqrt{\lambda}t)w(x_1, x_2)$ with

$$\Delta^2 w = \lambda w. \tag{3.3}$$

The simplest boundary conditions occur if the plate is clamped on its edges. Then

$$w = \partial w/\partial n = 0, \tag{3.4}$$

where $\partial w/\partial n$ represents the directional derivative of w normal to the boundary. If the boundary is restrained from moving vertically but no torque is applied, we find the so-called simply supported boundary conditions

$$w = 0,$$

$$M[w] \equiv -\nu\Delta w - (1-\nu)\left(\frac{\partial^2 w}{\partial x_1^2}n_1^2 + 2\frac{\partial^2 w}{\partial x_1 x_2}n_1 n_2 + \frac{\partial^2 w}{\partial x_2^2}n_2^2\right) = 0, \tag{3.5}$$

where ν is Poisson's ratio (an elastic constant) and n_1 and n_2 are the components of $\mathbf{n}$.

Finally, if no force or torque is applied, we have

$$M[w] = 0,$$

$$P[w] \equiv \frac{\partial}{\partial n}\Delta w + (1-\nu)\frac{\partial}{\partial t}\left(\frac{\partial^2 u}{\partial x_1^2}n_1 t_1 + \frac{\partial^2 u}{\partial x_1 \partial x_2}(n_1 t_2 + n_2 t_1) + \frac{\partial^2 u}{\partial x_2^2}n_1 t_2\right) = 0, \tag{3.6}$$

where **t** is a unit tangent vector and $\partial/\partial t$ represents the directional derivative in the direction of **t**.

We may, of course, have various of these conditions or various combinations of them on various parts of the boundary.

(c) *Buckling of a plate.* If a homogeneous isotropic plate is subjected to tangential forces, a certain stress field σ_{ij} is set up. The potential energy due to a normal displacement w is then found to be

$$(3.7)\quad \frac{1}{2}\int_D\int\left\{\nu(\Delta w)^2 + (1-\nu)\left(\frac{\partial^2 w}{\partial x_1^2} - \frac{\partial^2 w}{\partial x_2^2}\right)^2 + 2(1-\nu)\left(\frac{\partial^2 w}{\partial x_1 \partial x_2}\right)^2\right\} dx_1\, dx_2 + \frac{1}{2}\int_D\int \sum \sigma_{ij}\frac{\partial w}{\partial x_i}\frac{\partial w}{\partial x_j}\, dx_1\, dx_2.$$

We now suppose that the initial stress is proportional to a parameter $(-\lambda)$

$$\sigma_{ij} = -\lambda\tau_{ij}.$$

The buckling problem then asks for what value of λ the potential energy (3.7) ceases to be positive definite. This leads to the differential equation of buckling

$$(3.8)\quad \Delta^2 w = -\lambda \sum \tau_{ij}\frac{\partial^2 w}{\partial x_i \partial x_j}$$

together with one or more of the boundary conditions (3.4), (3.5), (3.6) depending on the physical condition of the boundary.

Some particularly simple cases are buckling under pure compression

$$(3.9)\quad \Delta^2 u = -\lambda \Delta u$$

and buckling under pure shear

$$(3.10)\quad \Delta^2 u = -\lambda\frac{\partial^2 u}{\partial x_1 \partial x_2}.$$

The relation between such linear problems and nonlinear buckling has been studied by Bauer and Reiss (1965), by Berger and Fife (1968), and by Knightly and Sather (1970), (1972).

(d) *The Bénard problem.* A viscous fluid in a vertical container is uniformly heated from below. The equations of motion are now the Navier–Stokes equations coupled with the equation of heat flow and the equation of state of the fluid. The side walls of the container are insulated so that no heat flows through them. The bottom is kept at temperature T_0 and the top at temperature $T_1 < T_0$.

An obvious solution is one in which the fluid remains at rest and heat is evenly conducted from bottom to top. However, the warmer fluid near the bottom is less dense than the cooler fluid on top. If the temperature difference is sufficiently great, the conduction solution becomes unstable and bifurcates into two other solutions in which fluid motion occurs.

The linearized problem for the velocity field **u**, the deviation θ from the equilibrium temperature distribution in the conduction solution, and the pressure p is described by equations which, in matrix notation, have the form

$$(3.11)\quad \begin{bmatrix} -\Delta & 0 & 0 & 0 & \dfrac{\partial}{\partial x_1} \\ 0 & -\Delta & 0 & 0 & \dfrac{\partial}{\partial x_2} \\ 0 & 0 & -\Delta & 0 & \dfrac{\partial}{\partial x_3} \\ 0 & 0 & 0 & -\Delta & 0 \\ \dfrac{\partial}{\partial x_1} & \dfrac{\partial}{\partial x_2} & \dfrac{\partial}{\partial x_3} & 0 & 0 \end{bmatrix} \begin{bmatrix} u_1 \\ u_2 \\ u_3 \\ \theta \\ p \end{bmatrix} = \lambda \begin{bmatrix} 0 & 0 & 0 & 0 & 0 \\ 0 & 0 & 0 & 0 & 0 \\ 0 & 0 & 0 & 1 & 0 \\ 0 & 0 & 1 & 0 & 0 \\ 0 & 0 & 0 & 0 & 0 \end{bmatrix} \begin{bmatrix} u_1 \\ u_2 \\ u_3 \\ \theta \\ p \end{bmatrix}.$$

The temperature difference $T_0 - T_1$ at which bifurcation can be expected to occur is proportional to the square of the lowest eigenvalue λ_1.

The boundary conditions are

$$(3.12)\qquad \mathbf{u} = \mathbf{0}, \qquad \partial\theta/\partial n = 0$$

on the side walls and

$$(3.13)\qquad \mathbf{u} = \mathbf{0}, \qquad \theta = 0$$

on both the bottom and top, provided there is a rigid lid on top. (If the top is free, we must change this last condition.)

The fact that increasing the temperature difference $T_0 - T_1$ above the value which corresponds to the eigenvalue λ_1 actually leads to a bifurcation has been established in various cases by Ukhovskii and Iudovich (1963), Iudovich (1966), Rabinowitz (1968), and Fife (1970).

CHAPTER 2

The Setting: Linear Vector Spaces and Their Properties

1. Linear vector spaces. We wish to have a uniform manner of describing the eigenvalue problems involving matrices, ordinary or partial differential operators, and even integro-differential operators which we have described in the preceding chapter.

We first define a linear vector space V over a field F which we shall always take to be the real or the complex numbers.

DEFINITION. A *linear vector space* V over a field F is a collection of elements $u, v, \cdots$, with the following properties:

(i) For every u in V and every α in F there is an element αu in V, and

$$\text{(a)}\quad \alpha(\beta u) = (\alpha\beta)u \quad \text{for all } u \in V \quad \text{and for all } \alpha, \beta \in F,$$

$$\text{(b)}\quad 1u = u \quad \text{for all } u \in V.$$

(ii) There is defined a mapping "+" which assigns to each pair of elements u, v of V an element $u + v$ of V with the properties

$$\text{(a)}\quad u + v = v + u \quad \text{for all } u, v \in V,$$

$$\text{(b)}\quad \alpha(u + v) = \alpha u + \alpha v \quad \text{for all } u, v \in V \quad \text{and for all } \alpha \in F,$$

$$\text{(c)}\quad (\alpha + \beta)u = \alpha u + \beta u \quad \text{for all } u \in V \quad \text{and for all } \alpha, \beta \in F.$$

(iii) There exists an element 0 of V with the property that

$$\text{(a)}\ u + 0 = u \quad \text{for all } u \in V,$$

and for $0 \in F$,

$$\text{(b)}\quad 0u = 0 \quad \text{for all } u \in V.$$

We note that by (i(b)) and (ii(c)),

$$u + (-1)u = 0 \quad \text{for all } u \in V.$$

We define

$$-u \equiv (-1)u.$$

Example 1.1. V_m is the collection of m-tuples $\mathbf{u} \sim (u_1, \cdots, u_m)$ of real (or complex) numbers. For real (or complex) α we define

$$\alpha(u_1, u_2, \cdots, u_m) = (\alpha u_1, \alpha u_2, \cdots, \alpha u_m)$$

and we let

$$(u_1, \cdots, u_m) + (v_1, \cdots, v_m) \equiv (u_1 + v_1, \cdots, u_m + v_m).$$

Example 1.2. Let V be the set of twice continuously differentiable functions $u(x)$ on the interval $[0, 1]$ which satisfy the relations $u(0) = 0$ and $u'(1) = 0$. We define

$$(\alpha u)(x) = \alpha \cdot u(x) \quad \text{for all } x \text{ in } [0, 1],$$

$$(u + v)(x) = u(x) + v(x) \quad \text{for all } x \text{ in } [0, 1].$$

Note that the boundary conditions in this example are homogeneous. If we required that $u(0) = 1$, the sum of two functions in V would not be in V, and hence V would not be a linear vector space.

Example 1.3. Let V be the set of m-tuples of functions of which the first $m - 1$ are continuously differentiable functions of $(x_1, \cdots, x_N)$ in a domain D in N-space, while the mth component is a continuous function defined on the boundary of D. The operations are defined by combining those of Examples 1.1 and 1.2.

2. Linear transformations. We make another important definition.

DEFINITION. A mapping T from a linear vector space V over F to another vector space W over F' which contains F is said to be a *linear transformation* if

$$T(\alpha u + \beta v) = \alpha T(u) + \beta T(v) \quad \text{for all } u, v \in V \quad \text{and} \quad \text{for all } \alpha, \beta \in F.$$

Note that the operation $\alpha u + \beta v$ is defined on the space V while $\alpha T(u) + \beta T(v)$ is defined on W.

There are natural definitions of addition of linear transformations from V to W and multiplication by an element of F':

$$(T_1 + T_2)(v) \equiv T_1(v) + T_2(v),$$

$$(\alpha T)(v) \equiv \alpha T(v).$$

It is easily verified that with these definitions the space $L(V, W)$ of linear transformations from V to W is a linear vector space over F'.

Example 2.1. Let V be the space V_m of Example 1.1 and let W be the space V_n of n-tuples. Let the image under the linear transformation T of the m-tuple $(0, \cdots, 0, 1, 0, \cdots, 0)$ with the 1 in the jth place be the element $(t_{1j}, t_{2j}, \cdots, t_{nj})$. Then because of the linearity of T,

$$\sum T(u_1, u_2, \cdots, u_m) = \left(\sum_{j=1}^{m} t_{1j}u_j, \sum_{j=1}^{m} t_{2j}u_j, \cdots, \sum_{j=1}^{m} t_{nj}u_j \right).$$

Conversely, for any $n \times m$ matrix (t_{ij}) this formula gives a linear transformation from V_m to V_n. Thus the space $L(V_m, V_n)$ is isomorphic to the space of $n \times m$ matrices with entries in F.

There are two interesting special cases of linear transformations.

DEFINITION. A linear transformation from V to itself is called a *linear operator*.

We now observe that we can identify the field F with the space V_1 of Example 2.1, so that F is a linear vector space.

DEFINITION. A linear transformation from the linear vector space V over F to F is called a *linear functional*. The space $L(V, F)$ is called the *dual space* of V.

We note that according to Example 2.1, $L(V_m, F)$ is isomorphic to the space of $1 \times m$ matrices, which, in turn, is isomorphic to V_m itself.

Example 2.2. If V is defined as in Example 1.2, then:

(a)

$$T(u) \equiv u'(x) + \int_x^1 y^2 u(y)\, dy$$

is a linear transformation from V to the space of once continuously differentiable functions on $[0, 1]$ which vanish at 1.

(b)

$$S(u) \equiv xu(x) - \int_0^x y^2 u(y)\, dy$$

is a linear operator, and

(c)

$$l(u) \equiv u''\left(\frac{1}{2}\right) + \int_0^1 y^2 u(y)\, dy$$

is a linear functional.

DEFINITION. A subset W of the elements of a linear vector space V with the property

$$\alpha u + \beta v \in W \quad \text{for all } u, v \in W \quad \text{and} \quad \text{for all } \alpha, \beta \in F$$

is called a *linear subspace* of V.

Example 2.3. If T is any linear transformation on V, then the set

$$N = \{u \in V \mid T(u) = 0\}$$

is a linear subspace of V. This set is called the *null space* of T.

(Note that the set of u with $T(u) = w \neq 0$, w fixed, is not closed under addition, and hence is not a linear subspace of V.)

DEFINITION. A set U of elements of a linear vector space V is said to be *linearly dependent* if there exist numbers $\alpha_1, \cdots, \alpha_k$ in F, not all of which are zero, and elements $u_1, \cdots, u_k$ in U so that

$$\alpha_1 u_1 + \cdots + \alpha_k u_k = 0.$$

Otherwise the set U of elements is said to be *linearly independent*.

If V contains a set of m linearly independent elements but every set of $m + 1$ elements is linearly dependent, we say that V has *dimension m*. It is easily seen that V is then isomorphic to the space V_m defined in Example 1.1. If for every positive integer m, V contains m linearly independent elements, we say that V is *infinite-dimensional*.

The spaces V defined in Examples 1.2 and 1.3 are infinite-dimensional.

3. Sesquilinear and quadratic functionals.

DEFINITION. A mapping $B(v, w)$ from the set $V \times W$ of ordered pairs of elements (v, w) with v in the linear vector space V over F and w in the linear vector space W over F to the field F is called a *sesquilinear functional* (or *form*) on $V \times W$ if, for each fixed w, $B(v, w)$ is a linear functional in V and for each fixed v the complex conjugate $\overline{B(v, w)}$ is a linear functional in W. If F is the field of real numbers, conjugation leaves F invariant, so that $B(u, v)$ is linear in both u and v. Such a real form is more commonly called a *bilinear functional* (or *form*).

We observe that by the above definition $B(u, v)$ is sesquilinear if and only if it satisfies the identity

$$B(\alpha u + \beta v, \gamma w + \delta x) = \alpha\bar{\gamma}B(u, w) + \alpha\bar{\delta}B(u, x) + \beta\bar{\gamma}B(v, w) + \beta\bar{\delta}B(v, x)$$

$$\text{for all } u, v \in V, \quad w, x \in W, \quad \alpha, \beta, \gamma, \delta \in F. \tag{3.1}$$

Example 3.1. Every sesquilinear functional on the space $V_m \times V_n$ with V_m and V_n defined as in Example 1.1 is of the form

$$B((u_1, \cdots, u_m), (v_1, \cdots, v_n)) = \sum_{j=1}^{m} \sum_{k=1}^{n} b_{jk} u_j \bar{v}_k.$$

Thus, B is defined by an $m \times n$ matrix (b_{ij}).

Example 3.2. If V is the vector space of Example 1.2,

$$B(u, v) \equiv \int_0^1 \int_0^1 (x^2 + y^3) \sin xy\, u(x)\overline{v''(y)}\, dx\, dy + u'(\tfrac{1}{2})\overline{v(1)}$$

is a sesquilinear functional on $V \times V$.

DEFINITION. A sesquilinear functional on $V \times V$ is said to be a *Hermitian functional* on V if

$$B(u, v) = \overline{B(v, u)}$$

for all u and v in V. If F is the field of real numbers, the bilinear form $B(u, v)$ is more commonly said to be *symmetric* if $B(u, v) = B(v, u)$.

Example 3.3. The sesquilinear functional in Example 3.1 is Hermitian if and only if $m = n$ and $b_{jk} = \bar{b}_{kj}$ for all j and k. The matrix (b_{ij}) is then said to be Hermitian.

DEFINITION. A mapping from a linear vector space V to the real numbers is said to be a *quadratic functional* (or *form*) if there exists a Hermitian (or, if V is over the field of real numbers, a symmetric) functional $B(u, v)$ on $V \times V$ in terms of which the mapping has the form $u \to B(u, u)$.

Note that if B is Hermitian, then $B(u, u) = \overline{B(u, u)}$ and hence $B(u, u)$ is real. The identity (3.1) implies that

$$B(u, v) = \tfrac{1}{4}[B(u + v, u + v) - B(u - v, u - v) + iB(u + iv, u + iv) - iB(u - iv, u - iv)], \tag{3.2}$$

so that the Hermitian form $B(u, v)$ can be recovered from a knowledge of the quadratic form $B(u, u)$ for all u in V. Thus, there is a one-to-one correspondence between Hermitian sesquilinear functionals and quadratic functionals.

When F is the field of real numbers, the symmetric bilinear form $B(u, v)$ that defines the quadratic functional $B(u, u)$ can be recovered from the formula

$$B(u, v) = \tfrac{1}{4}[B(u + v, u + v) - B(u - v, u - v)]. \tag{3.3}$$

Definition. A quadratic functional is said to be *positive definite* if $B(u, u)$ is positive for all u in V other than $u = 0$. (Clearly $B(0,0) = 0$.) It is said to be *positive semidefinite* if $B(u, u)$ is nonnegative for all u in V.

We state an important property of positive semidefinite quadratic functionals.

Theorem 5.1 (Schwarz's inequality). *If $B(u, v)$ is a Hermitian sesquilinear functional and the corresponding quadratic functional $B(u, u)$ is positive semidefinite, then for all u and v in V,*

$$|B(u, v)|^2 \leqq B(u, u)B(v, v). \tag{3.4}$$

Equality holds if and only if there is a nontrivial linear combination $\alpha u + \beta v$ for which $B(\alpha u + \beta v, \alpha u + \beta v) = 0$.

Proof. For any pair of elements u, v in V and any real θ, the quadratic polynomial

$$B(u + \xi e^{i\theta} v, u + \xi e^{i\theta} v) = B(u, u) + 2\xi \operatorname{Re}[e^{-i\theta} B(u, v)] + \xi^2 B(v, v) \tag{3.5}$$

in the real variable ξ is never negative. It follows that this polynomial cannot have two distinct real zeros. Therefore, if $B(v, v) \neq 0$, the discriminant must be nonpositive:

$$\{2 \operatorname{Re}[e^{-i\theta} B(u, v)]\}^2 - 4B(u, u)B(v, v) \leqq 0.$$

If $B(u, v) \neq 0$, we choose $\theta = \arg B(u, v)$ to obtain (3.4). If $B(u, v) = 0$, (3.4) is trivial. Finally if $B(v, v) = 0$, then we conclude from the fact that the polynomial (3.5) cannot become negative, that the coefficient of ξ must also vanish, so that $B(u, v) = 0$, and (3.4) is again satisfied.

If equality holds in (3.4) and $B(u, v) \neq 0$, then for $\theta = \arg B(u, v)$ the discriminant of the polynomial (3.5) vanishes and hence the polynomial has a real double zero. Thus for some ξ and θ, $B(u + \xi e^{i\theta} v, u + \xi e^{i\theta} v) = 0$. On the other hand, if $B(u, v) = 0$ equality in (3.4) implies that either $B(u, u) = 0$ or $B(v, v) = 0$. In any of these cases, the last statement of the theorem is verified.

If F happens to be the field of real numbers, the same proof works if θ is kept at 0 throughout.

Theorem 1 has the following immediate corollaries.

Corollary 1. *If $B(u, u)$ is positive semidefinite, then $B(v, v) = 0$ implies $B(u, v) = 0$ for all $u \in V$.*

Corollary 2. *If $B(u, u)$ is positive definite, equality in* (3.4) *holds if and only if u and v are linearly dependent.*

4. Hilbert space.

Definition. A linear vector space V on which there is defined a nonnegative real function $\|u\|$ with the properties

$$
(4.1)\qquad
\begin{aligned}
&\|u + v\| \leqq \|u\| + \|v\| \quad \text{for all } u, v \in V,\\
&\|\alpha u\| = |\alpha|\,\|u\| \quad \text{for all } u \in V, \quad \alpha \in F,\\
&\|u\| = 0 \Rightarrow u = 0,
\end{aligned}
$$

is called a *normed linear vector space*. The function $\|u\|$ is called the *norm* of u.

The inequality in (4.1) is called the *triangle inequality*. If we think of $\|u - v\|$ as the distance from u to v, this inequality states that the length of one side of a triangle is at most the sum of the lengths of the other two sides. In other words, the shortest distance between two points is a straight line.

We now observe that if $B(u, u)$ is any positive definite quadratic functional, then by Schwarz's inequality (3.4)

$$
\begin{aligned}
B(u + v, u + v) &= B(u, u) + 2\,\mathrm{Re}\, B(u, v) + B(v, v)\\
&\leqq [B(u, u)^{1/2} + B(v, v)^{1/2}]^2.
\end{aligned}
$$

Therefore,

$$
B(u + v, u + v)^{1/2} \leqq B(u, u)^{1/2} + B(v, v)^{1/2}.
$$

Moreover, $B(\alpha u, \alpha u)^{1/2} = |\alpha| B(u, u)^{1/2}$. Thus, the function $B(u, u)^{1/2}$ satisfies the axioms (4.1) for a norm.

DEFINITION. A normed linear vector space V on which the norm $\|u\|$ is equal to the square root of a positive definite quadratic functional $B(u, u)$ is called a *pre-Hilbert space*. The Hermitian functional $B(u, v)$ is called the *scalar product* of u and v.

Note that because of Schwarz's inequality we can define an angle θ between any two nonzero elements u and v of a pre-Hilbert space by the formula

$$
\cos\theta = \frac{B(u, v)}{B(u, u)^{1/2} B(v, v)^{1/2}}.
$$

In particular, we say that u and v are *orthogonal* if $B(u, v) = 0$.

A norm on a linear vector space V defines a *topology*. We say that the sequence v_n *converges to* v if $\|v - v_n\|$ approaches zero. We see from the triangle inequality that

$$
\|v_n - v_m\| \leqq \|v_n - v\| + \|v - v_m\|.
$$

Thus, a necessary condition for a sequence v_n to have a limit v in the normed linear space V is that

$$
\lim_{\substack{m \to \infty \\ n \to \infty}} \|v_n - v_m\| = 0.
$$

This is called the *Cauchy criterion*, and a sequence with this property is called a *Cauchy sequence*.

DEFINITION. A normed linear vector space in which every Cauchy sequence has a limit is said to be *complete*. Such a space is called a *Banach space*.

DEFINITION. A complete normed linear vector space with the property that the square of the norm $\|u\|^2$ is a positive definite quadratic functional $B(u, u)$ is called a *Hilbert space*. The corresponding Hermitian functional $B(u, v)$ is called the *scalar product* of u and v.

We shall now show how to complete a pre-Hilbert space to a Hilbert space.

DEFINITION. A linear functional l on a normed linear vector space is said to be *bounded* if there is a constant c so that

$$|l(u)| \leqq c\|u\| \quad \text{for all } u \in V.$$

The smallest admissible constant in this inequality is called the *norm* of l and is denoted by $\|l\|^*$. Note that for two linear functionals l_1 and l_2,

$$|l_1(u) + l_2(u)| \leqq (\|l_1\|^* + \|l_2\|^*)\|u\|.$$

and hence,

$$\|l_1 + l_2\|^* \leqq \|l_1\|^* + \|l_2\|^*.$$

Clearly $\|\alpha l\|^* = |\alpha|\,\|l\|^*$, and $\|l\|^* = 0$ implies that $l(u) = 0$ for all u, so that l is the zero functional. Thus the space V^* of bounded linear functionals on the normed linear vector space V is again a normed linear vector space. It is called the *adjoint space* of V.

THEOREM 4.1. *The adjoint space V^* of any normed linear vector space is complete.*

Proof. Let $\{l_n\}$ be a sequence of bounded linear functionals, and suppose that

$$\lim_{\substack{n\to\infty\\ m\to\infty}} \|l_n - l_m\|^* = 0.$$

For any fixed u in V we have

$$|l_n(u) - l_m(u)| \leqq \|l_n - l_m\|^*\|u\|,$$

which approaches zero. Thus $\{l_n(u)\}$ is a Cauchy sequence of complex numbers. Hence it has a limit, which we call $l(u)$:

$$l(u) = \lim_{n\to\infty} l_n(u).$$

Since each l_n is linear, we have

$$\begin{aligned} l(\alpha u + \beta v) &= \lim_{n\to\infty} l_n(\alpha u + \beta v) \\ &= \lim_{n\to\infty} [\alpha l_n(u) + \beta l_n(v)] \\ &= \alpha l(u) + \beta l(v), \end{aligned}$$

so that $l(u)$ is a linear functional on V.

For any fixed u in V,

$$\begin{aligned} |l(u) - l_n(u)| &\leqq |l(u) - l_m(u)| + |l_m(u) - l_n(u)| \\ &\leqq |l(u) - l_m(u)| + \|l_m - l_n\|\,\|u\|. \end{aligned}$$

Given $\varepsilon > 0$, we can make $\|l_m - l_n\| < \varepsilon$ by choosing $m \geqq n \geqq n_\varepsilon$. Thus

$$|l(u) - l_n(u)| \leqq |l(u) - l_m(u)| + \varepsilon\|u\|.$$

We now let $m \to \infty$ to see that for $n > n_\varepsilon$,

$$|l(u) - l_n(u)| \leqq \varepsilon\|u\| \quad \text{for all } u \text{ in } V.$$

That is,

$$\|l - l_n\|^* \leqq \varepsilon.$$

This shows both that l is a bounded linear functional and that

$$\lim_{n\to\infty} \|l - l_n\|^* = 0.$$

Thus each Cauchy sequence l_n converges to an element l of V^*.

THEOREM 4.2. *If V is a pre-Hilbert space, then V^* is a Hilbert space and V is isometrically equivalent to a dense linear subspace of V^*.*

Proof. To each v in V there corresponds a linear functional l_v defined by

$$l_v(u) \equiv B(u, v).$$

By Schwarz's inequality (3.4), l_v is bounded and

$$\|l_v\|^* \leqq B(v, v)^{1/2} = \|v\|.$$

Moreover,

$$\|l_v\|^*\|v\| \geqq |l_v(v)| = B(v, v) = \|v\|^2.$$

Therefore, $\|l_v\|^* = \|v\|$. Thus the mapping $v \to l_v$ from V to V^* is an isometry. Moreover, $l_{\alpha v + \beta w} = \bar{\alpha} l_v + \bar{\beta} l_w$. Such a mapping is said to be an anti-isomorphism.

We shall now show that the image of V under the mapping $v \to l_v$ is dense in V^*. Let l be any nonzero bounded linear functional on V. Then by definition

$$\|l\|^* = \sup_{u \in V} \frac{|l(u)|}{\|u\|}. \tag{4.2}$$

Therefore there exists a maximizing sequence u_n so that $|l(u_n)|/\|u_n\|$ increases to $\|l\|^*$. We let

$$v_n = \frac{\|l\|^{*2}}{l(u_n)} u_n$$

so that

$$\|v_n\| \searrow \|l\|^* \tag{4.3}$$

and

$$l(v_n) = \|l\|^{*2}. \tag{4.4}$$

We now recall that $\|u\|^2$ is a quadratic functional $B(u, u)$. Therefore the quantity

$$Q(u, u) \equiv \|l\|^{*2}\|u\|^2 - |l(u)|^2$$

is also a quadratic functional with the corresponding Hermitian functional

$$Q(u, v) \equiv \|l\|^{*2} B(u, v) - l(u)\overline{l(v)}.$$

By the definition of $\|l\|^*$, $Q(u, u)$ is positive semidefinite. Therefore we may apply Schwarz's inequality (3.4):

$$|Q(u, v_n)|^2 \leqq Q(v_n, v_n)Q(u, u). \tag{4.5}$$

By (4.4) we have for arbitrary u,

$$Q(u, v_n) = \|l\|^{*2}[l_{v_n}(u) - l(u)].$$

In particular, when $u = v_n$,

$$Q(v_n, v_n) = \|l\|^{*2}[\|v_n\|^2 - \|l\|^{*2}].$$

Thus (4.5) becomes

$$\begin{aligned}\|l\|^{*4}|l_{v_n}(u) - l(u)|^2 &\leqq \|l\|^{*2}[\|v_n\|^2 - \|l\|^{*2}][\|l\|^{*2}\|u\|^2 - l(u)^2] \\ &\leqq \|l\|^{*4}[\|v_n\|^2 - \|l\|^{*2}]\|u\|^2 \quad \text{for any } u \in V.\end{aligned}$$

Therefore,

$$\|l_{v_n} - l\|^{*2} \leqq \|v_n\|^2 - \|l\|^{*2},$$

and (4.3) shows that l_{v_n} converges to l. Thus the image of V under the mapping $v \to l_v$ is dense in V^*.

We now define the sesquilinear functional $B(l, \tilde{l})$ on V^* as follows:

$$B(l_v, l_w) \equiv \overline{B(v, w)}.$$

If l is the limit of a sequence l_{v_n} and $\tilde{l}$ is the limit of a sequence l_{w_m}, then

$$\begin{aligned}|B(l_{v_n}, l_{w_m}) - B(l_{v_p}, l_{w_q})| &= |B(l_{v_n} - l_{v_p}, l_{w_m}) + B(l_{v_p}, l_{w_m} - l_{w_q})| \\ &= |B(v_n - v_p, w_m) + B(v_p, w_m - w_q)| \\ &\leqq \|v_n - v_p\|\,\|w_m\| + \|v_p\|\,\|w_m - w_q\| \\ &= \|l_{v_n} - l_{v_p}\|\,\|l_{w_m}\| + \|l_{v_p}\|\,\|l_{w_m} - l_{w_q}\|.\end{aligned}$$

Since l_{v_n} and l_{w_m} are convergent sequences, the right-hand side goes to zero as n, m, p, and $q \to \infty$. It follows that the sequence $B(l_{v_n}, l_{w_m})$ has a limit, and we define

$$B(l, \tilde{l}) = \lim_{n,m \to \infty} B(l_{v_n}, l_{w_m}).$$

Clearly, $B(l, \tilde{l})$ is again Hermitian.

In particular,

$$\|l\|^{*2} = B(l, l),$$

and $B(l, l) = 0$ implies $l = 0$. By definition V^* is a Hilbert space.

As an immediate corollary we have the following.

COROLLARY. *A Hilbert space V is isometrically equivalent to its adjoint space V^*.*

Every bounded linear functional $l(u)$ can be written in the form $l(u) = B(u, v)$ for some v in V.

We remark that a real Hilbert space H can always be enlarged to a complex Hilbert space H_c in the following manner. To each element u of H we associate a new element iu. We then define addition and multiplication by scalars by means of the axioms of § 1. From the symmetric bilinear functional $B(u, v)$ on H we form the Hermitian sesquilinear functional

$$B(u_1 + iu_2, v_1 + iv_2) \equiv B(u_1, v_1) + B(u_2, v_2) + i[B(u_2, v_1) - B(u_1, v_2)].$$

In particular,

$$B(u_1 + iu_2, u_1 + iu_2) = B(u_1, u_1) + B(u_2, u_2).$$

Hence if $B(u, u)$ is positive definite on H, it is also positive definite on the complexification H_c.

5. The Sobolev spaces. Consider a domain D in N dimensions. We let V consist of the infinitely differentiable (real- or complex-valued) functions $u(x)$ defined on the closure of D. If D is unbounded, we also require each function $u(x)$ of V to vanish outside some bounded set, which may depend on u.

The quadratic functional

$$\|u\|^2_{H^0(D)} \equiv \int_D |u|^2\, dx \tag{5.1}$$

is clearly positive definite. It can be shown that every bounded linear functional on the pre-Hilbert space V with norm $\|u\|_{H^0(D)}$ can be written in the form

$$l(u) = \int_D u(x)\overline{v(x)}\, dx,$$

where $v(x)$ is a function which is Lebesgue measurable and for which

$$\int_D |v|^2\, dx$$

in the sense of Lebesgue is finite. Thus, the Hilbert space V^* is the space of Lebesgue square integrable functions on D. We can say that we have completed the space V to the Hilbert space $H^0(D)$ of functions for which the norm (5.1) with the integral taken in the sense of Lebesgue is finite. Every such function is the limit in the sense of the norm (5.1) of a sequence of functions in V. This space $H^0(D)$ is more usually denoted by $L_2(D)$.

The following norm is of great importance in the study of partial differential equations. For any nonnegative integer k we define

$$\|u\|^2_{H^k(D)} = \sum_{\alpha_1+\alpha_2+\cdots+\alpha_N \leqq k} \int_D \left| \frac{\partial^{\alpha_1+\cdots+\alpha_N} u}{\partial x_1^{\alpha_1} \cdots \partial x_N^{\alpha_N}} \right|^2 dx. \tag{5.2}$$

We introduce this norm on the space V defined above to create a pre-Hilbert space. The corresponding scalar product is

$$B(u, v) = \sum_{\alpha_1 + \cdots + \alpha_N \leq k} \int_D \frac{\partial^{\alpha_1 + \cdots + \alpha_N} u}{\partial x_1^{\alpha_1} \cdots \partial x_N^{\alpha_N}} \frac{\partial^{\alpha_1 + \cdots + \alpha_N} \bar{v}}{\partial x_1^{\alpha_1} \cdots \partial x_N^{\alpha_N}} \, dx. \tag{5.3}$$

By Theorem 4.2, any bounded linear functional is a limit of $B(u, v_n)$ with v_n a Cauchy sequence. Because of the definition (5.2) of the norm we see that then each partial derivative $\partial^{\alpha_1 + \cdots + \alpha_N} v / \partial x_1^{\alpha_1} \cdots \partial x_N^{\alpha_N}$ of order at most k of v_n forms a Cauchy sequence in the $H^0(D)$-norm. Hence it must have a square integrable limit $v_{\alpha_1 \cdots \alpha_N}(x)$:

$$\lim_{n \to \infty} \int_D \left| \frac{\partial^{\alpha_1 + \cdots + \alpha_N} v_n}{\partial x_1^{\alpha_1} \cdots \partial x_N^{\alpha_N}} - v_{\alpha_1 \cdots \alpha_N} \right|^2 dx = 0, \qquad \alpha_1 + \cdots + \alpha_N \leq k. \tag{5.4}$$

Taking the limit of $B(u, v_n)$, we see that any bounded linear functional can be expressed in the form

$$l(u) = \sum_{\alpha_1 + \cdots + \alpha_N \leq k} \int_D \frac{\partial^{\alpha_1 + \cdots + \alpha_N} u}{\partial x_1^{\alpha_1} \cdots \partial x_N^{\alpha_N}} \bar{v}_{\alpha_1 \cdots \alpha_N} \, dx, \tag{5.5}$$

where the functions $v_{\alpha_1 \cdots \alpha_N}$ are the limits in the sense (5.4) of derivatives of a sequence of functions v_n in V. If the limiting relation (5.4) holds, we define the functions $v_{\alpha_1 \cdots \alpha_N}$ to be *strong derivatives* of the limiting function $v_{00 \cdots 0}$. We denote this fact by

$$v_{\alpha_1 \cdots \alpha_N} = \frac{\partial^{\alpha_1 + \cdots + \alpha_N} v_{00 \cdots 0}}{\partial x_1^{\alpha_1} \cdots \partial x_N^{\alpha_N}}. \tag{5.6}$$

If we use this notation and write v for $v_{00 \cdots 0}$, we see from (5.5) that any bounded linear functional can be written in the form (5.3) where the derivatives on the right are strong derivatives. The completion of V, then, is the Hilbert space of square integrable functions having square integrable strong derivatives of all orders up to order k with the norm defined by (5.2). This space is called the *Sobolev space* $H^k(D)$.

We note that for $k > l$,

$$\|u\|_{H^k(D)} \geq \|u\|_{H^l(D)},$$

and hence that $H^k(D) \subset H^l(D)$.

CHAPTER 3

The Existence and Characterization of Eigenvalues

1. Formulation of the self-adjoint eigenvalue problem. We wish to consider the eigenvalue problem

$$Au = \lambda Bu, \tag{1.1}$$

where A and B are two given linear transformations from a linear vector space V to another linear vector space W which may or may not be the same as V. That is, we seek to find values of λ for which the equation (1.1) has solutions other than $u = 0$.

We shall treat only those eigenvalue problems for which the following hypotheses are satisfied.

HYPOTHESES 1.1. *There exists a sesquilinear functional* (v, w) *on* $V \times W$ *with the properties*:

(a) *If* $(v, w) = 0$ *for all* v *in* V, *then* $w = 0$.

(b) *The sesquilinear functionals* (u, Av) *and* (u, Bv) *are Hermitian*; *that is,*

$$(u, Av) = \overline{(v, Au)}, \qquad (u, Bv) = \overline{(v, Bu)} \quad \textit{for all } u, v \textit{ in } V.$$

(c) *The quadratic form* (u, Au) *is positive definite on* V, *and there is a constant* c *so that*

$$|(u, Bu)| \leqq c(u, Au). \tag{1.2}$$

We note that Hypothesis 1.1(c) implies that $\lambda = 0$ is not an eigenvalue of the problem (1.1). It is therefore possible, and it will turn out to be useful, to write the problem (1.1) in the form

$$Bu = \mu Au, \tag{1.3}$$

where $\mu = 1/\lambda$. It is possible that this problem has the eigenvalue $\mu = 0$ if B has a nontrivial null space. In this case, we simply say that $\lambda = \infty$ is an eigenvalue of problem (1.1).

Remark. Hypothesis (c) may be replaced by:

(c′) *There are real numbers* α, β, γ *and* δ *so that* $\alpha\delta - \beta\gamma \neq 0, \alpha(u, Au) + \beta(u, Bu)$ *is positive definite, and the one-sided inequality*

$$\gamma(u, Au) + \delta(u, Bu) \leqq \alpha(u, Au) + \beta(u, Bu)$$

holds.

In this case we replace the problem (1.1) by

$$(\alpha - \gamma)Au + (\beta - \delta)Bu = \mu[(2\alpha - \gamma)Au + (2\beta - \delta)Bu],$$

where

$$\mu = \frac{(\alpha - \gamma)\lambda + \beta - \delta}{(2\alpha - \gamma)\lambda + 2\beta - \delta}.$$

Example 1.1. Let $V = W = V_m$. Then there are $m \times m$ matrices (a_{kl}) and (b_{kl}) so that the problem (1.1) takes the form

$$\sum_{l=1}^{m} a_{kl}u_l = \lambda \sum_{l=1}^{m} b_{kl}u_l, \qquad k = 1, \cdots, m.$$

Since any sesquilinear functional (v, w) can be written in the form

$$(v, w) = \sum_{j,k=1}^{m} C_{jk}v_j\bar{w}_k,$$

Hypotheses 1.1 state that there should exist a matrix C_{jk} so that the matrix products

$$\sum_{k=1}^{m} C_{jk}\bar{a}_{kl} \quad \text{and} \quad \sum_{k=1}^{m} C_{jk}\bar{b}_{kl}$$

are Hermitian, and that the first of these products is positive definite. Hypothesis (c′) would mean that some linear combination of these two products is positive definite.

We remark that if $V = V_m$ and $W = V_n$, then for $n > m$ hypothesis (a) cannot hold, while for $n < m$ hypothesis (c) can never be satisfied. If either V or W is finite-dimensional, V and W must have the same dimension.

Example 1.2. Let D be a bounded domain in Euclidean n-space with smooth boundary $\partial D = \Sigma_1 \cup \Sigma_2$ where $\Sigma_1 \cap \Sigma_2$ is empty. Let V consist of functions $u(x)$ which are infinitely differentiable on the closure of D and which satisfy the boundary conditions

$$(1.4) \qquad \begin{aligned} u &= 0 \quad \text{on } \Sigma_1, \\ \partial u/\partial n + \rho(x)u &= 0 \quad \text{on } \Sigma_2, \end{aligned}$$

where $\partial u/\partial n$ is the outward normal derivative on Σ_2 and $\rho(x)$ is a given non-negative continuous function. Let W be the space of all functions which are infinitely differentiable on the closure of D.

Let A be minus the Laplace operator and let B be the identity, or rather the injection I from V to W. Then the eigenvalue problem (1.1) becomes

$$-\Delta u = \lambda u$$

with u subject to the boundary conditions (1.4).

We choose

$$(u, w) = \int_D u\bar{w}\,dx.$$

Then

$$(u, Bv) = \int_D u\bar{v}\,dx,$$

while by the divergence theorem

$$(u, Av) = \int_D \operatorname{grad} u \cdot \operatorname{grad} \bar{v}\,dx + \int_{\Sigma_2} \rho u\bar{v}\,dS.$$

Clearly Hypotheses 1.1(a) and 1.1(b) are satisfied. The fact that Hypothesis 1.1(c) also holds will follow from Theorems 2.5 and 2.8 of the next section.

We note that if we choose

$$(u, w) \equiv -\int_D \Delta u\bar{w}\,dx,$$

then

$$(u, Av) = \int_D \Delta u \Delta \bar{v}\,dx$$

and

$$(u, Bv) = \int_D \operatorname{grad} u \cdot \operatorname{grad} \bar{v}\,dx + \int_{\Sigma_2} \rho u\bar{v}\,dx$$

so that Hypotheses 1.1 are again satisfied. Thus the choice of the sesquilinear form (u, v) is by no means unique.

Example 1.3. We consider the equations (3.2) of Chap. 1 for the normal modes of a vibrating elastic solid which occupies a bounded domain D. For the sake of simplicity we consider the boundary conditions $\mathbf{u} = 0$.

Let V be the space of all real vector fields $\mathbf{v}$ which are defined and have bounded continuous partial derivatives of order two in D, and which vanish on the boundary of D. Let W be the space of all bounded continuous vector fields. We put $(\mathbf{v}, \mathbf{w}) = \int_D \mathbf{v} \cdot \mathbf{w}\,dx$. Then

$$(\mathbf{u}, A\mathbf{w}) = \sum_{i,j,k,l=1}^{3} \int_D A_{ijkl} \frac{\partial u_i}{\partial x_j} \frac{\partial w_k}{\partial x_l}\,dx$$

and

$$(\mathbf{u}, B\mathbf{w}) = \int_D \rho \mathbf{u} \cdot \mathbf{w}\,dx$$

for all $\mathbf{u}$ and $\mathbf{w}$ in V. Thus, Hypotheses 1.1(a) and 1.1(b) are satisfied. Hypothesis 1.1(c) follows from Theorems 2.5 and 2.8 of the next section, provided we make the additional hypothesis that for any symmetric matrix m_{ij} the inequality

$$\sum_{i,j,k,l=1}^{3} A_{ijkl} m_{ij} m_{kl} \geqq \mu \sum_{i,j=1}^{3} m_{ij}^2$$

holds with a positive constant μ. This inequality essentially states that the equilibrium state $\mathbf{u} = 0$ is stable.

Example 1.4. We consider the Bénard problem (3.11) of Chap. 1 with the boundary conditions (3.12) and (3.13) of Chap. 1.

If we let W be the space of five-tuples $\{u_1, u_2, u_3, \theta, p\}$, V the subspace of elements of W which satisfy the boundary conditions, and if we set

$$(\{v_1, v_2, v_3, \theta, p\}, \{w_1, w_2, w_3, \phi, q\}) = \int (\mathbf{v} \cdot \mathbf{w} + \theta\phi + pq)\, dx,$$

we find that we can satisfy Hypotheses 1.1(a) and 1.1(b), but not 1.1(c) or even 1.1(c′).

Instead, we let W_0 be the space of four-tuples $\{u_1, u_2, u_3, \theta\}$. We set

$$(\{v_1, v_2, v_3, \theta\}, \{w_1, w_2, w_3, \phi\}) = \int (\mathbf{v} \cdot \mathbf{w} + \theta\phi)\, dx.$$

Let W be the subspace of those elements $\{\mathbf{v}, \theta\}$ of W_0 for which $\operatorname{div} \mathbf{v} = 0$ and which satisfy the boundary conditions $\mathbf{v} \cdot \mathbf{n} = 0$. The orthogonal complement of W in W_0 with respect to the above scalar product consists of all elements of the form $\{\operatorname{grad} q, 0\}$. Moreover, any element of W_0 has a unique decomposition of the form

$$\{\mathbf{u}, \theta\} = \{\mathbf{v}, \theta\} + \{\operatorname{grad} q, 0\}$$

with $\{\mathbf{v}, \theta\}$ in W. We define the projection

$$P\{\mathbf{u}, \theta\} \equiv \{\mathbf{v}, \theta\},$$

and note that $P\{\mathbf{r}, s\} = 0$ if and only if $\mathbf{r}$ is a gradient and $s = 0$. We let V be the subspace of elements $\{\mathbf{v}, \theta\}$ of W with $\mathbf{v} = \partial\theta/\partial n = 0$ on the side walls and $\mathbf{v} = \theta = 0$ on the top and bottom.

We now let

$$A\{\mathbf{v}, \phi\} \equiv P\{-\Delta\mathbf{v}, -\Delta\phi\}$$

and

$$B\{\mathbf{v}, \phi\} \equiv P\{0, 0, \phi, v_3\}.$$

Then the equation

$$A\{\mathbf{u}, \theta\} = \lambda B\{\mathbf{u}, \theta\}$$

together with the condition $\{\mathbf{u}, \theta\} \in V$ is equivalent to the system (3.11)–(3.13) of Chap. 1.

We see that for $\{\mathbf{u}, \theta\}$ and $\{\mathbf{v}, \phi\}$ in V,

$$(\{\mathbf{u}, \theta\}, A\{\mathbf{v}, \phi\}) = \int \left[\sum_{i=1}^{3} \operatorname{grad} u_i \cdot \operatorname{grad} v_i + \operatorname{grad} \theta \cdot \operatorname{grad} \phi\right] dx$$

and

$$(\{\mathbf{u}, \theta\}, B\{\mathbf{v}, \phi\}) = \int (u_3 \phi + \theta v_3)\, dx.$$

It is easily verified that Hypotheses 1.1 are satisfied.

The above formulation is due to Sorokin (1953). An alternative reduction to a problem which satisfies Hypotheses 1.1 was given by Rabinowitz (1968).

We now define the Hermitian functionals

$$\mathscr{A}(u, v) = (u, Av), \qquad \mathscr{B}(u, v) = (u, Bv) \tag{1.5}$$

on V. By Hypothesis 1.1(c), $\mathscr{A}(u, u)$ is positive definite, so the space V with the norm $\mathscr{A}(u, u)^{1/2}$ is a pre-Hilbert space. We complete it to a Hilbert space $V_{\mathscr{A}}$ in which V is dense.

Example 1.5. Let A be an elliptic differential operator of the form

$$Au = \sum_{\substack{\alpha_1 + \cdots + \alpha_N \leqq M \\ \beta_1 + \cdots + \beta_N \leqq M}} \frac{\partial^{\alpha_1 + \cdots + \alpha_N}}{\partial x_1^{\alpha_1} \cdots \partial x_N^{\alpha_N}} \left[(-1)^{\alpha_1 + \cdots + \alpha_N} a_{\alpha_1 \cdots \alpha_N \beta_1 \cdots \beta_N}(x) \frac{\partial^{\beta_1 + \cdots + \beta_N} u}{\partial x_1^{\beta_1} \cdots \partial x_N^{\beta_N}} \right], \tag{1.6}$$

where the coefficient functions satisfy

$$a_{\beta_1 \cdots \beta_N \alpha_1 \cdots \alpha_N}(x) = \overline{a_{\alpha_1 \cdots \alpha_N \beta_1 \cdots \beta_N}(x)}. \tag{1.7}$$

Let D be a bounded N-dimensional domain with smooth boundary ∂D, and choose

$$(u, v) = \int_D u \bar{v}\, dx.$$

The boundary conditions which define the space V are assumed to make $\mathscr{A}(u, v)$ Hermitian. That is,

$$\int u A \bar{v}\, dx = \int \bar{v} A u\, dx \tag{1.8}$$

for all u and v in V. An integration by parts shows that for any smooth functions u and v

$$\begin{aligned} \mathscr{A}(u, v) &\equiv \int_D u A \bar{v}\, dx \\ &= \int_D \sum a_{\alpha_1 \cdots \alpha_N \beta_1 \cdots \beta_N} \frac{\partial^{\alpha_1 + \cdots + \alpha_N} u}{\partial x_1^{\alpha_1} \cdots \partial x_N^{\alpha_N}} \frac{\partial^{\beta_1 + \cdots + \beta_N} \bar{v}}{\partial x_1^{\beta_1} \cdots \partial x_N^{\beta_N}}\, dx + \int_{\partial D} \mathscr{R}(x; u, v)\, dS, \end{aligned} \tag{1.9}$$

where for each x on the boundary $\mathscr{R}(x; u, v)$ is a sesquilinear functional in the vector whose components are u and its partial derivatives up to order $M - 1$ and the vector whose components are v and its partial derivatives up to order $2M - 1$.

We shall assume that V consists of those smooth functions in D which satisfy M boundary conditions on ∂D, and that the fact that u and v satisfy these conditions, together with an application of the divergence theorem to the boundary integral, allows us to eliminate all derivatives of order higher than $M - 1$ from $\mathscr{R}(x; u, v)$ and to make $\mathscr{R}(x; u, v)$ Hermitian.

We further assume that A is coercive on V, in the sense that there is a constant c so that

$$\|u\|_{H^M(D)} \leqq c\mathscr{A}(u, u) \tag{1.10}$$

for all u in V. Under these assumptions, we see that the elements of $V_{\mathscr{A}}$ are the elements of the closure of V in $H^M(D)$.

We now observe that by Hypothesis 1.1(c) the quadratic functional $\mathscr{B}(u, u) + c\mathscr{A}(u, u)$ is positive semidefinite. Hence by Schwarz's inequality

$$\begin{aligned}|\mathscr{B}(u, v) + c\mathscr{A}(u, v)|^2 &\leqq [\mathscr{B}(u, u) + c\mathscr{A}(u, u)][\mathscr{B}(v, v) + c\mathscr{A}(v, v)]\\ &\leqq 4c^2\mathscr{A}(u, u)\mathscr{A}(v, v).\end{aligned}$$

Therefore,

$$|\mathscr{B}(u, v)| \leqq 3c\mathscr{A}(u, u)^{1/2}\mathscr{A}(v, v)^{1/2}. \tag{1.11}$$

It follows that for each fixed v, $\mathscr{B}(u, v)$ is a bounded linear functional on the pre-Hilbert space V. Therefore, there is an element Tv of the completion $v_{\mathscr{A}}$ such that

$$\mathscr{B}(u, v) = \mathscr{A}(u, Tv). \tag{1.12}$$

Moreover, we see from (1.11) that

$$\|Tv\| \leqq 3c\|v\|.$$

It is clear that T is a linear transformation from V to $V_{\mathscr{A}}$, and we have shown that it is bounded. Hence its definition can be extended by continuity to $V_{\mathscr{A}}$. We thus obtain a linear operator T on $V_{\mathscr{A}}$. Then $\mathscr{B}(u, v)$ is also defined on the completion $V_{\mathscr{A}}$ by means of (1.12).

We shall now relate the eigenvalue problem (1.3) on V with the problem

$$Tu = \mu u \tag{1.13}$$

on the completion $V_{\mathscr{A}}$.

THEOREM 1.1. *If u is an eigenvector corresponding to the eigenvalue μ of the problem* (1.3), *then it is also an eigenvector corresponding to the eigenvalue μ of the problem* (1.13). *Conversely, if an eigenvector u of* (1.13) *lies in V, then it is an eigenvector of* (1.3) *corresponding to the same eigenvalue.*

Proof. If u satisfies (1.3), form (v, Bu) with an arbitrary v in V and use (1.3) to find

$$\mathscr{B}(v, u) = \bar{\mu}\mathscr{A}(v, u) \quad \text{for all } v \in V.$$

Then by the definition (1.12),

$$\mathscr{A}(v, Tu - \mu u) = 0$$

for all v in V. Since V is dense in $V_{\mathscr{A}}$ we can take limits of v in V to find that this equation still holds for a v in $V_{\mathscr{A}}$. In particular, we may choose $v = Tu - \mu u$ to find that $\mathscr{A}(Tu - \mu u, Tu - \mu u) = 0$. Since $\mathscr{A}(u, u)$ is positive definite, this implies (1.13).

Suppose now that (1.13) is satisfied with u in V. Then

$$\mathscr{A}(v, Tu) = \bar{\mu}\mathscr{A}(v, u) \quad \text{for all } v \text{ in } V_{\mathscr{A}}.$$

By (1.5) and (1.12) we see that

$$(v, Bu - \mu Au) = 0 \quad \text{for all } v \text{ in } V.$$

Hence, by Hypothesis 1.1(a), $Bu - \mu Au = 0$, which completes the proof.

This theorem tells us that solving the eigenvalue problem (1.13) produces all the eigenvalues of the original problem (1.1). However, it does not preclude the possibility that (1.13) has eigenvectors in the completion $V_{\mathscr{A}}$ which are not in the original space V.

In the important case where A is of the form (1.6), B is of order lower than $2M$, and the assumptions of Example 1.5 are satisfied, it can be shown (see, e.g., Fichera [1965]) that if the coefficients of A and B and the boundary of D are smooth, then every solution of the problem (1.13) is smooth. Hence it lies in V if V is chosen to be the space of functions having $2M$ continuous derivatives and satisfying the boundary conditions.

We shall show how to approximate the eigenvalues and eigenvectors of T. However, it should be kept in mind that the operator T cannot in general be found explicitly. Note that formally $T = A^{-1}B$. In Example 1.2 where $A = -\Delta$, $B = I$, T is the solution operator for the problem

$$-\Delta u = f$$

with the boundary conditions (1.4). That is, the solution of this problem is $u = Tf$.

Our approximation methods must involve only computable quantities such as $\mathscr{A}(u, v)$ and $\mathscr{B}(u, v)$. Since $\mathscr{B}(u, v)$ is Hermitian, T satisfies the identity

$$\mathscr{A}(u, Tv) = \mathscr{A}(Tu, v).$$

Such an operator is said to be *self-adjoint* on the Hilbert space $V_{\mathscr{A}}$.

We prove a simple consequence of self-adjointness.

THEOREM 1.2. *Every eigenvalue of the problem* (1.13), *and hence also of the problems* (1.1) *and* (1.3), *is real.*

Proof. If (1.13) is satisfied.

$$\mathscr{A}(u, Tu) = \bar{\mu}\mathscr{A}(u, u)$$

or

$$\mathscr{B}(u, u) = \bar{\mu}\mathscr{A}(u, u).$$

Taking imaginary parts, we see that

$$\operatorname{Im}(\mu)\mathscr{A}(u, u) = 0.$$

Since $u \neq 0$, this implies that $\operatorname{Im}(\mu) = 0$, which completes the proof.

We observe that this theorem requires not only the symmetry hypothesis, Hypothesis 1.1(b), but also the positive definiteness (1.1(c) or 1.1(c′)) of at least some linear combination of (u, Au) and (u, Bu). For example, the eigenvalue problem

$$\begin{pmatrix} 0 & 1 \\ 1 & 0 \end{pmatrix}\begin{pmatrix} x_1 \\ x_2 \end{pmatrix} = \lambda \begin{pmatrix} 1 & 0 \\ 0 & -1 \end{pmatrix}\begin{pmatrix} x_1 \\ x_2 \end{pmatrix},$$

whose coefficient matrices are real and symmetric, has the eigenvalues $\lambda = \pm i$.

2. The eigenvalues as successive maxima. We recall that the ratio $\mathscr{B}(v, v)/\mathscr{A}(v, v)$ is by hypothesis bounded on the Hilbert space $V_{\mathscr{A}}$. We define

$$\mu_1 = \underset{v \in V}{\text{l.u.b.}} \frac{\mathscr{B}(v, v)}{\mathscr{A}(v, v)}, \tag{2.1}$$

where the least upper bound is taken with respect to all nonzero vectors in V. We first prove the following result.

THEOREM 2.1. *If there exists a nonzero vector u_1 in $V_{\mathscr{A}}$ for which $\mathscr{B}(u_1, u_1) = \mu_1 \mathscr{A}(u_1, u_1)$, then*

$$Tu_1 = \mu_1 u_1. \tag{2.2}$$

Proof. By the definition (2.1) the quadratic functional $\mu_1\mathscr{A}(v, v) - \mathscr{B}(v, v)$ is positive semidefinite. Hence by Schwarz's inequality,

$$|\mu_1\mathscr{A}(v, u_1) - \mathscr{B}(v, u_1)|^2 \leqq [\mu_1\mathscr{A}(v, v) - \mathscr{B}(v, v)][\mu_1\mathscr{A}(u_1, u_1) - \mathscr{B}(u_1, u_1)] = 0$$

for all v in $V_{\mathscr{A}}$. Putting $v = \mu_1 u_1 - Tu_1$, we see that

$$\mathscr{A}(\mu_1 u_1 - Tu_1, \mu_1 u_1 - Tu_1) = 0,$$

and hence $Tu_1 - \mu_1 u_1 = 0$. Note that we can always multiply u_1 by a constant to make $\mathscr{A}(u_1, u_1) = 1$.

If the supremum μ_1 is attained, we thus have one eigenvalue. We then wish to find other eigenvalues. To facilitate this process we prove the following basic result.

THEOREM 2.2. *Let μ and $\tilde{\mu}$ be two different eigenvalues of the problem* (1.13) *and let u and $\tilde{u}$ be the corresponding eigenvectors. Then $\mathscr{A}(u, \tilde{u}) = 0$. That is, u and $\tilde{u}$ are orthogonal.*

Proof. Take the scalar product in $V_{\mathscr{A}}$ of $\tilde{u}$ with the eigenvalue equation for u and vice versa:

$$\mathscr{A}(\tilde{u}, Tu) = \mu\mathscr{A}(\tilde{u}, u), \qquad \mathscr{A}(u, T\tilde{u}) = \tilde{\mu}\mathscr{A}(u, \tilde{u}).$$

Subtract the complex conjugate of the second equation from the first, and recall that both bilinear forms are Hermitian and that $\tilde{\mu}$ is real to find

$$(\mu - \tilde{\mu})\mathscr{A}(u, \tilde{u}) = 0.$$

Since $\mu \neq \tilde{\mu}$, this implies $\mathscr{A}(u, \tilde{u}) = 0$, which completes the proof.

If we already have the eigenvector u_1 and we wish to find other eigenvectors, we need only look at the subspace V_1 of elements of $V_{\mathscr{A}}$ which are orthogonal to u_1. It is easily verified that V_1 is again a Hilbert space.

We now define

$$\mu_2 = \underset{\mathscr{A}(v,u_1)=0}{\text{l.u.b.}} \frac{\mathscr{B}(v, v)}{\mathscr{A}(v, v)}.$$

As in Theorem 2.1 we find that if there is a u_2 in V_1 for which $(u_2, Tu_2) = \mu_2(u_2, u_2)$, then u_2 satisfies the equation

$$\mathscr{A}(v, Tu_2 - \mu_2 u_2) = 0 \tag{2.3}$$

for all v in V_1. Now

$$\begin{aligned} \mathscr{A}(Tu_2 - \mu_2 u_2, u_1) &= \mathscr{A}(Tu_2, u_1) - \mu_2 \mathscr{A}(u_2, u_1) \\ &= \mathscr{A}(u_2, Tu_1) - \mu_2 \mathscr{A}(u_2, u_1) \\ &= (\mu_1 - \mu_2)\mathscr{A}(u_2, u_1) \\ &= 0 \end{aligned}$$

since u_2 is in V_1. Hence (2.3) is valid for all v in V, and we may choose $v = Tu_2 - \mu_2 u_2$ to see that $Tu_2 = \mu_2 u_2$.

We can continue in this manner to prove the following result.

THEOREM 2.3. *The recursive definition,*

(a)

$$\mu_n = \underset{\substack{\mathscr{A}(v,u_1)=0 \\ \mathscr{A}(v,u_2)=0 \\ \vdots \\ \mathscr{A}(v,u_{n-1})=0 \\ \mathscr{A}(v,v)=1}}{\text{l.u.b.}} \mathscr{B}(v, v), \tag{2.4}$$

(b) *if there is an element u in $V_{\mathscr{A}}$ which satisfies the conditions*

$$\mathscr{A}(u, u_1) = \mathscr{A}(u, u_2) = \cdots = \mathscr{A}(u, u_{n-1}) = 0,$$

$$\mathscr{A}(u, u) = 1, \qquad \mathscr{B}(u, u) = \mu_n,$$

define $u_n = u$,

leads to a nonincreasing sequence of eigenvalues $\mu_1 \geqq \mu_2 \geqq \cdots$ and eigenvectors $u_1, u_2, \cdots$ which terminates if and only if for some n the least upper bound in (2.4) *is not attained or V has dimension $n - 1$.*

Note that the same number may occur several times in the sequence of eigenvalues with different corresponding eigenvectors. We call the number of times it occurs the *multiplicity* of the eigenvalue. If $\mu_n = \mu_{n+1} = \cdots = \mu_{n+k-1}$, then any linear combination of $u_n, \cdots, u_{n+k-1}$ is again an eigenvector corresponding to this eigenvalue.

Remark. If V has finite dimension N, μ_n with $n \leq N$ is defined in (2.4) as the maximum of a continuous function over a bounded set in Euclidean N-space. It is therefore always attained. On the other hand, since the conditions $\mathscr{A}(v, u_1) = \cdots = \mathscr{A}(v, u_N) = 0$ imply $v = 0$, μ_n is not defined for $n > N$.

If $V = V_N$ and (u, v) is the usual Euclidean scalar product, Hypotheses 1.1 require that A and B be represented by Hermitian matrices and that A be positive definite. If we define Q to be the $N \times N$ matrix whose nth column is the eigenvector u_n which corresponds to the eigenvalue μ_n of the problem

$$Bu = \mu A u,$$

we see that

$$BQ = AQM,$$

where

$$M = \begin{pmatrix} \mu_1 & & & \\ & \mu_2 & 0 & \\ & 0 & \ddots & \\ & & & \mu_N \end{pmatrix}.$$

If Q^* denotes the complex conjugate of the transpose of the matrix Q, we see from (2.4) that

$$Q^*AQ = I,$$

the unit matrix, and hence that

$$Q^*BQ = M,$$

a diagonal matrix. Thus we have the following corollary of Theorem 2.3.

COROLLARY. *If A and B are two Hermitian $N \times N$ matrices and A is positive definite, there is a nonsingular matrix Q with the properties that Q^*AQ is the unit matrix and Q^*BQ is a real diagonal matrix. The diagonal entries of Q^*BQ are the eigenvalues of the problem $Bu = \mu Au$.*

We note that the problem

$$\begin{pmatrix} 1 & 0 \\ 0 & 0 \end{pmatrix} \begin{pmatrix} x_1 \\ x_2 \end{pmatrix} = \mu \begin{pmatrix} 0 & 1 \\ 1 & 0 \end{pmatrix} \begin{pmatrix} x_1 \\ x_2 \end{pmatrix}$$

has the single eigenvalue $\mu = 0$ and the single eigenvector $\{0, 1\}$, even though the coefficient matrices are Hermitian. This shows that Hypothesis 1.1(c) or 1.1(c′) is essential for the above corollary.

To show that the definition of Theorem 2.3 gives us all the eigenvalues down to the point where it terminates, we prove the following result.

THEOREM 2.4. *There are no eigenvalues of T above μ_1, and if μ is an eigenvalue that satisfies $\mu > \mu_n$ for some n, then $\mu = \mu_k$ for some $k < n$, and the corresponding eigenvector is a linear combination of those eigenvectors u_i which correspond to the eigenvalue μ_k.*

Proof. We first observe that if $Tu = \mu u$, then $\mathscr{B}(u, u) = \mu\mathscr{A}(u, u)$. Hence by definition of μ_1, $\mu \leqq \mu_1$. Now if $\mu > \mu_n$ for some n, then there is a $k \leqq n - 1$ such that

$$\mu_{k+1} < \mu \leqq \mu_k.$$

If $\mu < \mu_k$ we see from Theorem 2.2 that $\mathscr{A}(u, u_1) = \cdots = \mathscr{A}(u, u_k) = 0$ and hence by the definition in Theorem 2.3 that $\mu \leqq \mu_{k+1}$, which gives a contradiction. Therefore $\mu = \mu_k$.

If u is not a linear combination of those u_i which correspond to μ_k, we can produce a linear combination of u and these u_i which is orthogonal to $u_1, \cdots, u_k$ and which is again an eigenvector corresponding to μ. But this would again make $\mu \leqq \mu_{k+1}$, and thus give a contradiction.

It is, of course, of interest to have a sufficient condition for the least upper bound to be attained. The following is related to a criterion of Friedrichs (1948).

THEOREM 2.5. *Suppose there exist a positive number η and finitely many linear functionals $l_1, \cdots, l_k$ on V with the property that the conditions*

$$l_j(v) = 0, \quad j = 1, \cdots, k \quad \text{and} \quad v \in V \tag{2.5}$$

imply that

$$\mathscr{B}(v, v) \leqq (\mu_n - \eta)\mathscr{A}(v, v), \tag{2.6}$$

where μ_n is the least upper bound defined in (2.4). *Then this least upper bound is attained, so that μ_n is an eigenvalue.*

Moreover, there is an integer l which satisfies $n \leqq l \leqq k$ with the property that for $n \leqq j \leqq l$ the supremum μ_j is equal to μ_n and is attained, and $\mu_{l+1} < \mu_n$.

Proof. For the sake of simplicity, we shall present the proof for the case $n = 1$. The proof for the more general case can be obtained by replacing the space V by the orthogonal complement in V of $u_1, u_2, \cdots, u_{n-1}$. Assume, then, that the conditions (2.5) imply that

$$\mathscr{B}(v, v) \leqq (\mu_1 - \eta)\mathscr{A}(v, v)$$

for all v in V, where

$$\mu_1 = \underset{\mathscr{A}(v,v)=1}{\text{l.u.b.}} \mathscr{B}(v, v).$$

By the definition of least upper bound, there is a maximizing sequence w_n such that

$$\mathscr{A}(w_n, w_n) = 1, \qquad \mathscr{B}(w_n, w_n) \nearrow \mu_1.$$

Suppose at first that the functionals $l_1, \cdots, l_k$ are all bounded. Then since $\mathscr{A}(w_n, w_n) = 1$, the sequences $l_1(w_n), \cdots, l_k(w_n)$ are all bounded. It follows that there is a subsequence w_n' of w_n on which the sequences $l_j(w_n')$ converge. We may assume without loss of generality that the linear functionals l_j are linearly independent in V. (Otherwise we simply throw out some of the l_j.) This fact implies that there are elements $q_1, \cdots, q_k$ in V so that

$$l_i(q_j) = \delta_{ij} = \begin{cases} 1 & \text{if } i = j, \\ 0 & \text{if } i \neq j. \end{cases} \tag{2.7}$$

We now define

$$p_n = w'_n - \sum_{j=1}^{k} l_j(w'_n) q_j$$

so that $l_1(p_n) = l_2(p_n) = \cdots = l_k(p_n) = 0$. By hypothesis, then,

$$\mathscr{B}(p_n - p_m, p_n - p_m) \leqq (\mu_1 - \eta)\mathscr{A}(p_n - p_m, p_n - p_m). \tag{2.8}$$

We apply the triangle inequality for the positive semidefinite functional $\mu_1 \mathscr{A}(v, v) - \mathscr{B}(v, v)$ to the element

$$p_n - p_m = w'_n - w'_m + \sum_{j=1}^{k} [l_j(w'_m) - l_j(w'_n)] q_j$$

to find that

$$\begin{aligned} &[\mu_1 \mathscr{A}(p_n - p_m, p_n - p_m) - \mathscr{B}(p_n - p_m, p_n - p_m)]^{1/2} \\ &\quad \leqq [\mu_1 \mathscr{A}(w'_n, w'_n) - \mathscr{B}(w'_n, w'_n)]^{1/2} + [\mu_1 \mathscr{A}(w'_m, w'_m) - \mathscr{B}(w'_m, w'_m)]^{1/2} \\ &\quad\quad + \sum_{j=1}^{k} |l_j(w'_n) - l_j(w'_m)| [\mu_1 \mathscr{A}(q_j, q_j) - \mathscr{B}(q_j, q_j)]^{1/2}. \end{aligned} \tag{2.9}$$

Combining (2.8) and (2.9), we have

$$\begin{aligned} \eta^{1/2} \mathscr{A}(p_n - p_m, p_n - p_m)^{1/2} \leqq{}& [\mu_1 \mathscr{A}(w'_n, w'_n) - \mathscr{B}(w'_n, w'_n)]^{1/2} \\ &+ [\mu_1 \mathscr{A}(w'_m, w'_m) - \mathscr{B}(w'_m, w'_m)]^{1/2} \\ &+ \sum_{j=1}^{k} |l_j(w'_n) - l_j(w'_m)| [\mu_1 \mathscr{A}(q_j, q_j) - \mathscr{B}(q_j, q_j)]^{1/2}. \end{aligned}$$

Since w'_n is a subsequence of a maximizing sequence, the first two terms approach zero. Since the sequences $l_j(w'_n)$ converge, the sum also goes to zero. It follows that p_n is a Cauchy sequence. Since

$$\|w'_n - w'_m\| \leqq \|p_n - p_m\| + \sum_{j=1}^{k} |l_j(w'_n) - l_j(w'_m)| \cdot \|q_j\|,$$

w'_n is also a Cauchy sequence. Therefore it has a limit in $V_{\mathscr{A}}$ which we call u_1. Since $\mathscr{A}(w'_n, w'_n) = 1$, we find that $\mathscr{A}(u_1, u_1) = 1$. Also,

$$\begin{aligned} [\mu_1 \mathscr{A}(u_1, u_1) - \mathscr{B}(u_1, u_1)]^{1/2} \leqq{}& [\mu_1 \mathscr{A}(u_1 - w'_n, u_1 - w'_n) - \mathscr{B}(u_1 - w'_n, u_1 - w'_n)]^{1/2} \\ &+ [\mu_1 \mathscr{A}(w'_n, w'_n) - \mathscr{B}(w'_n, w'_n)]^{1/2}. \end{aligned}$$

Since both terms on the right approach zero, we find that $\mu_1\mathscr{A}(u_1, u_1) = \mathscr{B}(u_1, u_1)$ and hence by Theorem 2.1 that $Tu_1 = \mu_1 u_1$.

This argument shows that if the l_i in the hypothesis are all bounded, the supremum μ_n is attained by an eigenvector u_n. Moreover, if $\mu_j = \mu_n$ for $j > n$, the same argument shows that there are orthonormal eigenvectors $u_1, \cdots, u_j$. If $j > k$, there is a linear combination u of these eigenvectors which satisfies the conditions (2.5). A computation shows that $\mathscr{B}(u, u) \geqq \mu_n \mathscr{A}(u, u)$, which contradicts (2.6). Thus $j \leqq k$ and Theorem 2.5 is proved when the l_j are all bounded.

If the l_j are not bounded, Theorem 2.5 is an immediate consequence of the above argument and the following lemma.

LEMMA 2.1. *Let the linear functions $l_1, \cdots, l_k$ on the dense linear subspace V of the Hilbert space $V_{\mathscr{A}}$ have the property that $l_1(v) = \cdots = l_k(v) = 0, v \in V$ implies*

$$\mathscr{B}(v, v) \leqq \tilde{c}\mathscr{A}(v, v). \tag{2.10}$$

Then there exist some linear combinations $\bar{l}_1, \cdots, \bar{l}_{k'}$ of $l_1, \cdots, l_k$ which are bounded linear functionals and which have the same property with the same constant $\tilde{c}$.

Proof. Suppose that the functionals $l_1, \cdots, l_k$ are not all bounded. Then there is a sequence r_n in V for which

$$\sum_{j=1}^{n} |l_j(r_n)|^2 = 1, \qquad \lim_{n\to\infty} \mathscr{A}(r_n, r_n) = 0.$$

Since each sequence $l_1(r_n), \cdots, l_k(r_n)$ is bounded, we can choose a subsequence r_n' of r_n so that the sequences $l_j(r_n')$ converge:

$$\lim_{n\to\infty} l_j(r_n') = \alpha_j, \qquad \sum_{j=1}^{k} |\alpha_j|^2 = 1.$$

We define the new elements

$$s_n = r_n' - \sum_{j=1}^{k} [l_j(r_n') - \alpha_j] q_j,$$

where the q_j have the property (2.7). Then

$$l_i(s_n) = \alpha_i, \qquad i = 1, \cdots, k,$$

and, because $\mathscr{A}(r_n', r_n')$ and $l_j(r_n') - \alpha_j$ approach zero,

$$\lim_{n\to\infty} \mathscr{A}(s_n, s_n) = 0.$$

We now choose $k - 1$ linearly independent k-tuples $b_{\nu j}$ in such a way that

$$\sum_{j=1}^{k} b_{\nu j}\alpha_j = 0, \qquad \nu = 1, \cdots, k - 1.$$

Consider any element v in V which satisfies the $k - 1$ linear conditions

$$\sum_{j=1}^{k} b_{\nu j} l_j(v) = 0, \qquad \nu = 1, \cdots, k - 1. \tag{2.11}$$

Since the rows of the matrix $b_{\nu j}$ are linearly independent, this system of equations implies that

$$l_j(v) = \gamma\alpha_j = \gamma l_j(s_n)$$

for some constant γ. Hence $l_1(v - \gamma s_n) = \cdots = l_n(v - \gamma s_n) = 0$. According to the hypothesis of this lemma

$$\frac{\mathscr{B}(v - \gamma s_n, v - \gamma s_n)}{\mathscr{A}(v - \gamma s_n, v - \gamma s_n)} \leqq \tilde{c}.$$

Letting $n \to \infty$ and recalling that $\mathscr{A}(s_n, s_n) \to 0$ we see that the $k - 1$ conditions (2.11) already imply the inequality (2.10).

If the $k - 1$ linear functionals $\sum b_{\nu j} l_j$ are not bounded, we go through the same procedure and continue until we obtain a (possibly empty) set of bounded linear functionals $\bar{l}_1, \cdots, \bar{l}_{k'}$ which are linear combinations of $l_1, \cdots, l_{k'}$ and which have the property that the conditions

$$\bar{l}_1(v) = \cdots = \bar{l}_{k'}(v) = 0$$

imply (2.10).

DEFINITION. The *discrete spectrum* of T is the set of eigenvalues of T. The *essential* (or *continuous*) *spectrum* of T is the set of those μ for which there exists a sequence v_l with the properties

$$\tag{2.12} \begin{aligned} \mathscr{A}(v_l, v_m) &= \delta_{lm}, \\ \mathscr{B}(v_l, v_m) &= 0 \quad \text{for } l \neq m, \\ \lim_{l \to \infty} \mathscr{B}(v_l, v_l) &= \mu. \end{aligned}$$

The union of the discrete spectrum and the essential spectrum is called the *spectrum* of T.

It is clear from (2.12) that the spectrum of T consists of real numbers μ.

We now relate the nonattainment of μ_n to the essential spectrum.

THEOREM 2.6. *If there is no element u_n for which the least upper bound μ_n in* (2.4) *is attained, or if μ_n is an eigenvalue of infinite multiplicity, then μ_n is the largest member of the essential spectrum of T.*

Conversely, if μ_n as defined by (2.4) *belongs to the essential spectrum, then either μ_n is an eigenvalue of infinite multiplicity, or there is an $l \geqq n$ for which $\mu_l = \mu_n$ and the greatest lower bound μ_l in* (2.4) *is not attained.*

Proof. If μ_n is an eigenvalue of infinite multiplicity, the sequence of eigenvectors $u_1, u_2, \cdots$ has the properties (2.12) with $\mu = \mu_n$.

If μ_n is not attained, choose any v_1 for which $\mathscr{A}(v_1, v_1) = 1$, $\mathscr{A}(v_1, u_1) = \cdots = \mathscr{A}(v_1, u_{n-1}) = 0$, and $\mathscr{B}(v_1, v_1) > \mu_n - 1$. Since μ_n is not attained, Theorem 2.5 shows that there is a v_2 with $\mathscr{A}(v_2, v_2) = 1$ which satisfies the linear conditions $\mathscr{A}(v_2, u_1) = \cdots = \mathscr{A}(v_2, u_{n-1}) = \mathscr{A}(v_2, v_1) = \mathscr{B}(v_2, v_1) = 0$ and for which $\mathscr{B}(v_2, v_1) > \mu_n - 1/2$. By the same argument, we see that if we already have $v_1, \cdots, v_j$, there is a v_{j+1} with $\mathscr{A}(v_{j+1}, v_{j+1}) = 1$ which satisfies the linear conditions

$$\mathscr{A}(v_{j+1}, u_1) = \cdots = \mathscr{A}(v_{j+1}, v_1) = \cdots = \mathscr{A}(v_{j+1}, v_j) = \mathscr{B}(v_{j+1}, v_1)$$
$$= \cdots = \mathscr{B}(v_{j+1}, v_j) = 0$$

and for which $\mathscr{B}(v_{j+1}, v_{j+1}) > \mu_n - (1/j + 1)$. Since $\mathscr{B}(v_l, v_l) \leqq \mu_n$, the sequence $\{v_l\}$ has the properties (2.12) with $\mu = \mu_n$, so that μ_n is in the essential spectrum.

If μ is any point of the essential spectrum, there is a sequence v_l with the properties (2.12). For any n for which μ_n is defined by (2.4), let $u_1, \cdots, u_{n-1}$ be the eigenvectors defined by (2.4). For each l we define v_l' to be a linear combination of $v_l, \cdots, v_{l+n}$ which is orthogonal to $u_1, \cdots, u_{n-1}$. It is easily seen that $\mathscr{B}(v_l', v_l')/\mathscr{A}(v_l', v_l')$ converges to μ. Therefore $\mu_n \geqq \mu$ for all n. That is, a point μ of the essential spectrum cannot lie above any of the eigenvalues defined by (2.4). This proves both that if μ_n is in the essential spectrum it must be its largest member and the last part of Theorem 2.6.

The sequence $\{v_l\}$ constructed in the proof of the first part of Theorem 2.6 behaves almost like an infinite orthonormal set of eigenvectors which correspond to the eigenvalue μ_n. For this reason we define $\mu_n = \mu_{n+1} = \mu_{n+2} = \cdots$ when μ_n is not attained so that μ_j for $j > n$ is not defined by (2.4).

Definition. The ratio $\mathscr{B}(u, u)/\mathscr{A}(u, u)$ is called the *Rayleigh quotient*. The values $\mu_1 \geqq \mu_2 \geqq \cdots$ defined by (2.4) with the convention that if μ_n is not attained, $\mu_n = \mu_{n+1} = \mu_{n+2} = \cdots$, are called the *eigenvalues of the Rayleigh quotient* $\mathscr{B}(u, u)/\mathscr{A}(u, u)$. They are defined whenever $\mathscr{A}(u, u)$ is positive definite and the quotient $\mathscr{B}(u, u)/\mathscr{A}(u, u)$ is bounded above.

In order to use Theorem 2.5 as a criterion to establish the existence of eigenvalues, we need a way of generating linear functionals with the needed properties. The most important result along these lines is the following.

Theorem 2.7 (Poincaré's inequality). *Let $v(x_1, \cdots, x_N)$ be a square integrable function with square integrable first partial derivatives on the N-dimensional cube K_a of side a ($v \in H^1(K_a)$). Then*

$$\int_{K_a} |v|^2\, dx \leqq a^{-N}\left|\int_{K_a} v\, dx\right|^2 + \tfrac{1}{2}Na^2 \int_{K_a} |\operatorname{grad} v|^2\, dx. \tag{2.13}$$

Proof. For the sake of simplicity, we shall prove the theorem in two dimensions. The ideas are the same in one dimension or in more dimensions.

We first assume that v is continuously differentiable. We introduce Cartesian coordinates in which K_a is the set $0 \leqq x_i \leqq a$, $i = 1, 2$. Choose two points (x_1, x_2) and (y_1, y_2) on the square. Clearly

$$v(y_1, y_2) - v(x_1, x_2) = \int_{x_1}^{y_1} \frac{\partial v}{\partial x_1}(\xi, x_2)\, d\xi + \int_{x_2}^{y_2} \frac{\partial v}{\partial x_2}(y_1, \eta)\, d\eta.$$

Applying Schwarz's inequality to the bilinear functional

$$\int_{x_1}^{y_1} \phi_1(\xi, x_2)\bar{\psi}_1(\xi, x_2)\, d\xi + \int_{x_2}^{y_2} \phi_2(y_1, \eta)\bar{\psi}_2(y_1, \eta)\, d\eta$$

with $\psi_1 = \psi_2 = 1$, we find that

$$\begin{aligned} |v(y_1, y_2) - v(x_1, x_2)|^2 &\leqq \{|y_1 - x_1| + |y_2 - x_2|\}\left\{\int_{x_1}^{y_1} \left|\frac{\partial v}{\partial x_1}(\xi, x_2)\right|^2 d\xi \right. \\ &\quad \left. + \int_{x_2}^{y_2} \left|\frac{\partial v}{\partial x_2}(y_1, \eta)\right|^2 d\eta\right\} \\ &\leqq 2a\left\{\int_0^a \left|\frac{\partial v}{\partial x_1}(\xi, x_2)\right|^2 d\xi + \int_0^a \left|\frac{\partial v}{\partial x_2}(y_1, \eta)\right|^2 d\eta\right\}. \end{aligned}$$

We square out the left side, and then integrate both sides with respect to x_1, x_2, y_1, and y_2 from 0 to a. In this way we find that

$$\begin{aligned} a^2 \int_{K_a} |v(y_1, y_2)|^2\, dy_1\, dy_2 &- \int_{K_a} v(y_1, y_2)\, dy_1\, dy_2 \int_{K_a} \overline{v(x_1, x_2)}\, dx_1\, dx_2 \\ &- \int_{K_a} \overline{v(y_1, y_2)}\, dy_1\, dy_2 \int_{K_a} v(x_1, x_2)\, dx_1\, dx_2 + a^2 \int_{K_a} |v(x_1, x_2)|^2\, dx_1\, dx_2 \\ \leqq 2a^4 \int_{K_a} &\left(\left|\frac{\partial v}{\partial x_1}\right|^2 + \left|\frac{\partial v}{\partial x_2}\right|^2\right) dx_1\, dx_2. \end{aligned}$$

If we observe that the variable of integration y can be replaced by x, we obtain (2.13) for $N = 2$.

By a limiting process we can extend this result to strongly differentiable functions, which completes the proof.

We remark that by eigenvalue techniques we can replace the constant $\frac{1}{2}N$ in (2.13) by π^{-2}. We can apply this result in two ways.

DEFINITION. A square integrable function v with square integrable strong first derivatives on a domain D is said to *vanish on the boundary of D* if it is the limit in the $H^1(D)$ norm ($\|u\|_{H^1}^2 = \int_D (|\text{grad}\, u|^2 + u^2)\, dx$) of a sequence of functions each of which vanishes outside some closed bounded (i.e., compact) subset of D. The space of these functions is a Hilbert space which is a subspace of $H^1(D)$. This space is denoted by $\mathring{H}^1(D)$.

THEOREM 2.8 (Friedrichs' inequality). *Let D be a domain which is contained in a cube K_b of side b. Then for any positive ε there is a finite set of linear functionals $l_1, \cdots, l_k$ so that if $v \in \mathring{H}^1(D)$, the conditions $l_1(v) = 0, \cdots, l_k(v) = 0$ imply that*

$$\int_D |v|^2\, dx \leqq \varepsilon \int_D |\text{grad}\, v|^2\, dx. \tag{2.14}$$

Proof. A function v in $\mathring{H}^1(D)$, extended as zero outside D, is clearly in $H^1(K_b)$. Choose an integer m so that $2m^2\varepsilon \geqq Nb^2$, and divide the sides of the cube into m

equal parts. This divides K_b into m^N cubes $K^{(1)}, K^{(2)}, \cdots, K^{(m^N)}$ of side b/m. Let

$$l^{(\nu)}(v) = \int_{K^{(\nu)}} v\,dx.$$

Then according to Theorem 2.7, $l^{(\nu)}(v) = 0$ implies that

$$\int_{K^{(\nu)}} |v|^2\,dx \leqq \frac{N}{2}\left(\frac{b}{m}\right)^2 \int_{K^{(\nu)}} |\text{grad } v|^2\,dx$$

$$\leqq \varepsilon \int_{K^{(\nu)}} |\text{grad } v|^2\,dx.$$

Summing these inequalities over all cubes gives (2.14).

If we wish to obtain an analogous result for functions which do not vanish on the boundary, we must assume that the boundary is somewhat smooth.

THEOREM 2.9. *Let D be a bounded domain with the property that a neighborhood $\mathcal{N}$ of its boundary can be covered by finitely many relatively open sets $S_1, \cdots, S_m$ such that in each set S_i there is a function $\phi_i(x)$ with the properties*

(i) *ϕ_i is continuously differentiable in S_i,*

(ii) *$\phi_i = 0$ and $|\text{grad } \phi_i| > 0$ on $S_i \cap \partial D$,*

(iii) *$\phi_i > 0$ on $S_i \cap D$.*

Then for any $\varepsilon > 0$ one can construct a finite number of linear functionals $l_1, \cdots, l_k$ so that

$$l_i(u) = 0, \qquad i = 1, \cdots, k,$$

implies

$$\int_D u^2\,dx \leqq \varepsilon \int_D |\text{grad } u|^2\,dx. \tag{2.15}$$

Proof. In the neighborhood S_1 introduce a new coordinate system $\xi_1, \cdots, \xi_N$ with $\xi_1 = \phi_1(x)$ and so that the Jacobian $\partial(\xi_1, \cdots, \xi_N)/\partial(x_1, \cdots, x_N)$ is bounded and bounded away from 0 in some subset S_1' with the property that $S_1' \cup S_2 \cup \cdots \cup S_m$ is a neighborhood of ∂D. Do the same for $S_2, \cdots, S_m$, so that $S_1' \cup \cdots \cup S_m'$ is a neighborhood of ∂D and on S_i' the boundary has the equation $\xi_1 = 0$. Choose an η so that the sets $0 \leqq \xi_1 \leqq \eta$ are in all the S_i' and let S_i'' be the subset of S_i' where this inequality holds. Since the Jacobian $\partial(\xi_1, \cdots, \xi_m)/\partial(x_1, \cdots, x_m)$ is bounded away from zero, there is a constant c so that

$$\int_{S_i''} |u|^2\,dx \leqq c \int_{S_i''} |u|^2\,d\xi, \tag{2.16}$$

$$\int_{S_i''} |\text{grad}_x\, u|^2\,dx \geqq c^{-1} \int_{S_i''} |\text{grad}_\xi\, u|^2\,d\xi.$$

Now cover the set $S_1'' - (S_2'' \cup \cdots \cup S_m'')$ by a union S_1''' of adjacent cubes of side η/n in the ξ-coordinates. The integer n is chosen large enough that this set can be so covered, and that for some ε_1, the condition that $\int u\,d\xi = 0$ on each cube implies that the integral of u^2 is less than ε_1 times the integral of $|\text{grad}_\xi u|^2$. Adding these inequalities, and using (2.16), we obtain the inequality

$$\int_{S_1'''} |u|^2\,dx \leqq \varepsilon_1 c^2 \int_{S_1'''} |\text{grad } u|^2\,dx.$$

We now do the same on S_2', then S_3', and so forth to obtain

$$\int_{S_i'''} |u|^2\,dx \leqq \varepsilon_1 c^2 \int_{S_i'''} |\text{grad } u|^2\,dx. \tag{2.17}$$

We now cover $D - (S_1'' \cup S_2'' \cup \cdots \cup S_m'')$ with a union D''' of adjacent cubes in D of sufficiently small size. Then if the integral of u over these cubes vanishes, we find

$$\int_{D'''} |u|^2\,dx \leqq \varepsilon_1 \int_{D'''} |\text{grad } u|^2\,dx. \tag{2.18}$$

We now add the inequalities (2.17) and (2.18). Taking into account the fact that the domains S_i''' and D''' may overlap so that the integral of $|\text{grad } u|^2$ over the same set may occur several times, we see that if the integral of u over all the ξ-cubes of the S_i''' and the x-cubes of D''' vanishes, then

$$\int_D |u|^2\,dx \leqq \varepsilon_1(mc^2 + 1)\int_D |\text{grad } u|^2\,dx.$$

Choosing $\varepsilon_1 = \varepsilon/(mc^2 + 1)$, we obtain the statement of the theorem.

COROLLARY. *Under the conditions of Theorem 2.9, given any $\varepsilon > 0$ and any integers p and q with $q > p$, we can construct finitely many linear functionals $l_1, \cdots, l_k$ so that $l_1(u) = \cdots = l_k(u) = 0$ implies*

$$\sum_{\alpha_1+\cdots+\alpha_N=p} \int_D \left|\frac{\partial^{\alpha_1+\cdots+\alpha_N} u}{\partial x_1^{\alpha_1}\cdots\partial x_N^{\alpha_N}}\right|^2 dx \leqq \varepsilon \sum_{\beta_1+\cdots+\beta_N=q} \int_D \left|\frac{\partial^{\beta_1+\cdots+\beta_N} u}{\partial x_1^{\beta_1}\cdots\partial x_N^{\beta_N}}\right|^2 dx.$$

Proof. Apply Theorem 2.9 to each of the derivatives appearing on the left to obtain the inequality with $q = p + 1$. Then apply the same theorem to the $(p + 1)$st derivatives to obtain the inequality with $q = p + 2$. Continue in this fashion to the desired q.

Example 2.1. Consider the Schroedinger equation

$$-\Delta u + Vu = \lambda u \tag{2.19}$$

in all of Euclidean N-space. We seek solutions u which are square integrable. We suppose that

$$\liminf_{|x|\to\infty} V(x) = 0$$

and that for some positive constant α,

$$V(x) \geqq -\alpha.$$

In order to fit this problem into the framework of our theory, we first rewrite it as

$$-\Delta u + (V + \alpha + 1)u = \tilde{\lambda} u,$$

where $\tilde{\lambda} = \lambda + \alpha + 1$. Then with

$$(u, w) = \int u\bar{w}\, dx$$

we have

$$\mathscr{A}(u, v) = \int [\operatorname{grad} u \cdot \operatorname{grad} \bar{v} + (V + \alpha + 1)u\bar{v}]\, dx,$$

$$\mathscr{B}(u, v) = \int u\bar{v}\, dx.$$

For any $\varepsilon > 0$ we can choose a radius R so large that

$$V > -\varepsilon \quad \text{for } |x| > R.$$

By Theorem 2.9 we can find a finite set of linear conditions $l_1(u) = \cdots = l_k(u) = 0$ which imply that

$$\int_{|x| \leqq R} |u|^2\, dx \leqq \varepsilon \int_{|x| \leqq R} |\operatorname{grad} u|^2\, dx.$$

But then

$$\mathscr{A}(u, u) \geqq \int_{|x| > R} (-\varepsilon + \alpha + 1)|u|^2\, dx + \int_{|x| < R} \left(1 + \frac{1}{\varepsilon}\right)|u|^2\, dx$$

$$\geqq (\alpha + 1 - \varepsilon)\mathscr{B}(u, u)$$

if $\varepsilon \leqq 1/\alpha$. Thus, we can make

$$\frac{\mathscr{B}(u, u)}{\mathscr{A}(u, u)} \leqq \frac{1}{\alpha + 1 - \varepsilon}$$

for any $\varepsilon > 0$. By Theorems 2.5 and 2.6 the set $\mu > 1/(\alpha + 1)$ contains no continuous spectrum. Note that $\lambda = \tilde{\lambda} - \alpha - 1 = 1/\mu - \alpha - 1$, so that $\mu > 1/(\alpha + 1)$ becomes $\lambda < 0$. The inequality

$$\frac{\mathscr{B}(u, u)}{\mathscr{A}(u, u)} > \frac{1}{\alpha + 1}$$

is equivalent to

$$\int [|\operatorname{grad} u|^2 + V|u|^2]\, dx < 0.$$

It follows from Theorem 2.5 that if the last inequality is satisfied for some u, then (2.19) has a negative eigenvalue. Any negative spectrum is discrete.

Example 2.2. Consider again the problem

$$-\Delta u = \lambda u \quad \text{in } D,$$
$$u = 0 \quad \text{on } \Sigma_1,$$
$$\partial u/\partial n + \rho u = 0 \quad \text{on } \Sigma_2 = \partial D - \Sigma_1, \quad \rho \geqq 0.$$

Putting

$$(u, w) = \int u\bar{w}\, dx$$

we have

$$\mathscr{A}(u, v) = \int_D \operatorname{grad} u \cdot \operatorname{grad} \bar{v}\, dx + \int_{\Sigma_2} \rho u\bar{v}\, dS, \tag{2.20}$$
$$\mathscr{B}(u, v) = \int u\bar{v}\, dx.$$

Suppose that D satisfies the conditions of Theorem 2.9. Then for any $\varepsilon > 0$ we can find linear conditions $l_1(u) = \cdots = l_k(u) = 0$ so that

$$\mathscr{B}(u, u) \leqq \varepsilon \mathscr{A}(u, u).$$

By Theorem 2.5 the sequence of eigenvalues μ_n cannot terminate for any positive μ_n. Since $\mathscr{B}(u, u) \geqq 0$, the whole spectrum is discrete except for $\mu = 0$.

We note that the same kind of argument works for the problem

$$-\Delta u = \lambda x_1 u$$

with the same boundary conditions. Here $\mathscr{A}(u, v)$ is again given by (2.20), but

$$\mathscr{B}(u, v) = \int x_1 u\bar{v}\, dx \tag{2.21}$$

so that $\mathscr{B}(u, u)$ is no longer positive. We can still say, however, that any $\mu_n > 0$ cannot be in the continuous spectrum. Moreover, we can define the successive eigenvalues $\mu_1^{(-)} \geqq \mu_2^{(-)} \geqq \cdots$ for the problem $-\mathscr{B}(u, u)/\mathscr{A}(u, u)$ in the same way. Thus, we find that T has only discrete spectrum for $\mu < 0$ as well as for $\mu > 0$. The only point of continuous spectrum is $\mu = 0$.

3. Complete continuity and Fourier series. We generalize the property of the preceding example to the following definition.

DEFINITION. The quadratic functional $\mathscr{B}(u, u)$ is said to be *completely continuous* with respect to $\mathscr{A}(u, u)$ if for any $\varepsilon > 0$ there is a finite set of linear functionals $l_1, \cdots, l_k$ so that $l_1(u) = l_2(u) = \cdots = l_k(u) = 0$ implies

$$|\mathscr{B}(u, u)| \leqq \varepsilon \mathscr{A}(u, u).$$

We see from Theorem 2.5 applied to both the operators T and $-T$ that if $\mathscr{B}(u, u)$ is completely continuous with respect to $\mathscr{A}(u, u)$, then T has no continuous spectrum except possibly at $\mu = 0$.

In fact, if V is infinite-dimensional, the definition (2.4) applied to $\mathscr{B}(u, u)$ and $-\mathscr{B}(u, u)$ gives a nonincreasing sequence of eigenvalues $\mu_1 \geqq \mu_2 \geqq \cdots$ and a nondecreasing sequence of eigenvalues $-\mu_1^{(-)} \leqq -\mu_2^{(-)} \leqq \cdots$, both of which converge to zero. Corresponding to these eigenvalues we have two sequences of eigenvectors u_n and $u_n^{(-)}$ which are all orthogonal and have norm 1.

Corresponding to such an orthonormal sequence there is a Fourier series. Let v be any element of V. We define

$$v_\varepsilon \equiv v - \sum_{\mu_n \geq \varepsilon} \mathscr{A}(v, u_n)u_n - \sum_{\mu_n^{(-)} \geq \varepsilon} \mathscr{A}(v, u_n^{(-)})u_n^{(-)}. \tag{3.1}$$

We see from the orthonormality of the u_n and $u_n^{(-)}$, that

$$\mathscr{A}(v_\varepsilon, v_\varepsilon) = \mathscr{A}(v, v) - \sum_{\mu_n \geqq \varepsilon} |\mathscr{A}(v, u_n)|^2 - \sum_{\mu_n^{(-)} \geqq \varepsilon} |\mathscr{A}(v, u_n^{(-)})|^2.$$

Since $\mathscr{A}(v_\varepsilon, v_\varepsilon) \geqq 0$, this implies

$$\sum_{\mu_n \geqq \varepsilon} |\mathscr{A}(v, u_n)|^2 + \sum_{\mu_n^{(-)} \geqq \varepsilon} |\mathscr{A}(v, u_n^{(-)})|^2 \leqq \mathscr{A}(v, v),$$

which is called *Bessel's inequality*. It shows in particular that the sums of the squares of the Fourier coefficients converge. We also see that if $\varepsilon_1 > \varepsilon_2 > 0$,

$$\mathscr{A}(v_{\varepsilon_2} - v_{\varepsilon_1}, v_{\varepsilon_2} - v_{\varepsilon_1}) = \sum_{\varepsilon_2 \leqq \mu_n < \varepsilon_1} |\mathscr{A}(v, u_n)|^2 + \sum_{\varepsilon_2 \leqq \mu_n^{(-)} < \varepsilon_1} |\mathscr{A}(v, u_n^{(-)})|^2.$$

Since the right-hand side is a sum of cuts of convergent series, it approaches zero as ε_1 and ε_2 approach zero.

We conclude that v_ε is a Cauchy sequence. Therefore it has a limit $\tilde{v}$ in $V_{\mathscr{A}}$ as $\varepsilon \to 0$.

It follows from the definition (3.1) that

$$\mathscr{A}(v_\varepsilon, u_n) = \mathscr{A}(v_\varepsilon, u_n^{(-)}) = 0 \quad \text{for } \mu_n \geqq \varepsilon, \quad \mu_n^{(-)} \geqq \varepsilon.$$

Letting $\varepsilon \to 0$, we see that

$$\mathscr{A}(\tilde{v}, u_n) = \mathscr{A}(\tilde{v}, u_n^{(-)}) = 0 \quad \text{for all } n \quad \text{with } \mu_n \neq 0 \quad \text{or} \quad \mu_n^{(-)} \neq 0.$$

Then also

$$\mathscr{B}(\tilde{v}, u_n) = \mathscr{B}(\tilde{v}, u_n^{(-)}) = 0 \quad \text{for all } n \quad \text{with } \mu_n \neq 0 \quad \text{or} \quad \mu_n^{(-)} \neq 0.$$

Finally, we see from (2.4) that on the whole space

$$\tilde{W} = \{\tilde{w} | \mathscr{A}(\tilde{w}, u_n) = \mathscr{A}(\tilde{w}, u_n^{(-)}) = 0 \text{ for all } n \text{ with } \mu_n \neq 0 \text{ or } \mu_n^{(-)} \neq 0\}$$

we have

$$\mathscr{B}(\tilde{w}, \tilde{w}) = 0.$$

Thus on $\tilde{W}$, $\mathscr{B}(\tilde{w}, \tilde{w})$ is a very simple positive semidefinite quadratic functional. It follows by Schwarz's inequality that

$$|\mathscr{B}(\tilde{v}, \tilde{w})|^2 \leqq \mathscr{B}(\tilde{v}, \tilde{v})\mathscr{B}(\tilde{w}, \tilde{w}) = 0.$$

Therefore $\mathscr{B}(\tilde{v}, \tilde{w}) = 0$ for all $\tilde{v}$ and $\tilde{w}$ in $\tilde{W}$.

We have shown that any v in $V_{\mathscr{A}}$ may be decomposed into the sum

$$v = \tilde{v} + \sum \mathscr{A}(v, u_n)u_n + \sum \mathscr{A}(v, u_n^{(-)})u_n^{(-)}, \qquad \tilde{v} \in \tilde{W}. \tag{3.2}$$

Now take another arbitrary element w of $V_{\mathscr{A}}$ and decompose it in the same manner:

$$w = \tilde{w} + \sum \mathscr{A}(w, u_n)u_n + \sum \mathscr{A}(v, u_n^{(-)})u_n^{(-)}. \tag{3.3}$$

We easily find that

$$\mathscr{B}(\tilde{v}, w) = \mathscr{B}(\tilde{v}, \tilde{w}) = 0.$$

Thus $\tilde{v}$ in the decomposition (3.2) has the property

$$\mathscr{B}(\tilde{v}, w) = 0 \quad \text{for all } w \in V_{\mathscr{A}}.$$

This means that $T\tilde{v} = 0$.

In other words, $\tilde{v}$ is an eigenvector of T corresponding to the eigenvalue $\mu = 0$. Thus, we see that (3.2) is a Fourier series expansion in the eigenvectors of T. We easily can show that if v and w have the expansions (3.2) and (3.3),

$$\begin{aligned} \mathscr{A}(v, w) &= \mathscr{A}(\tilde{v}, \tilde{w}) + \sum_{n=1}^{\infty} \mathscr{A}(v, u_n)\overline{\mathscr{A}(w, u_n)} + \sum_{n=1}^{\infty} \mathscr{A}(v, u_n^{(-)})\overline{\mathscr{A}(w, u_n^{(-)})}, \\ \mathscr{B}(v, w) &= \sum_{n=1}^{\infty} \mu_n \mathscr{A}(v, u_n)\overline{\mathscr{A}(w, u_n)} - \sum_{n=1}^{\infty} \mu_n^{(-)} \mathscr{A}(v, u_n^{(-)})\overline{\mathscr{A}(w, u_n^{(-)})}. \end{aligned} \tag{3.4}$$

These are *Parseval's equations.*

The null space of T may have finite or infinite dimension. The simplest case occurs when it is empty; that is, when $\mathscr{B}(\tilde{v}, w) = 0$ for all w in V implies that $\tilde{v} = 0$. This is the case in Example 2.2. In this case the function $\tilde{v}$ is zero, and may be left out of (3.2) and (3.4). We summarize our results in the following theorem.

THEOREM 3.1. *If $\mathscr{B}(u, u)$ is completely continuous with respect to $\mathscr{A}(u, u)$, the definition* (2.4) *applied to $\mathscr{B}(u, u)$ and $-\mathscr{B}(u, u)$ gives two sequences of eigenvalues $\mu_1 \geqq \mu_2 \geqq \cdots$ and $-\mu_1^{(-)} \leqq -\mu_2^{(-)} \leqq \cdots$, both of which converge to zero. The corresponding eigenvectors u_n and $u_n^{(-)}$ have the property that the Fourier expansion theorem* (3.2) *holds when convergence is defined in the norm $\mathscr{A}(u, u)^{1/2}$ and $\tilde{v}$ is a solution of $T\tilde{v} = 0$. The Parseval equations* (3.4) *are also valid.*

Remark. The corollary to Theorem 2.9 shows that if the domain D satisfies the conditions of Theorem 2.9, if $\mathscr{A}(u, u)$ is greater than a positive constant times the integral of the sum of the squares of all partial derivatives of u up to some order q and if $\mathscr{B}(u, u)$ involves only lower derivatives, then $\mathscr{B}(u, u)$ is completely continuous with respect to $\mathscr{A}(u, u)$. However, this sort of result is not valid if the domain D is allowed to become unbounded.

Consider, for example, a domain D in 2-space which contains the infinite strip $|x_2| \leqq a$. Let $\phi(x_1, x_2)$ be any infinitely differentiable function which vanishes outside the disk $x_1^2 + x_2^2 \leqq a^2$. Let

$$\phi_n(x_1, x_2) \equiv \phi(x_1 + 2na, x_2), \qquad n = 1, 2, \cdots.$$

Let

$$\mathscr{A}(\phi, \phi) \equiv \int_D |\text{grad } \phi|^2 \, dx_1 \, dx_2 = 1$$

and put

$$\mathscr{B}(\phi, \phi) \equiv \int_D |\phi|^2 \, dx_1 \, dx_2 = \alpha > 0.$$

Then clearly

$$\mathscr{A}(\phi_n, \phi_m) = \delta_{nm}, \qquad \mathscr{B}(\phi_n, \phi_m) = \alpha\delta_{nm}.$$

Given any finite set of linear functionals $l_1, l_2, \cdots, l_k$, we can find a nontrivial linear combination

$$u = \sum_{n=1}^{k+1} c_n \phi_n$$

for which $l_1(u) = \cdots = l_k(u) = 0$. But

$$\mathscr{B}(u, u)/\mathscr{A}(u, u) = \alpha.$$

Thus no matter how many constraints we impose the ratio $\mathscr{B}(u, u)/\mathscr{A}(u, u)$ remains α, and hence $\mathscr{B}$ is not completely continuous with respect to $\mathscr{A}$.

Finally, we observe that from the expansion (3.2) and Parseval's equations (3.4) it follows that

$$Tv = \sum_{n=1}^{\infty} \mu_n \mathscr{A}(v, u_n) u_n - \sum_{n=1}^{\infty} \mu_n^{(-)} \mathscr{A}(v, u_n^{(-)}) u_n^{(-)}.$$

From this expansion and the fact that the μ_n and $\mu_n^{(-)}$ approach zero it is easy to see that T has the following two properties:

(a) If $\{v_\alpha\}$ is any bounded sequence, there is a subsequence $\{v'_\alpha\}$ with the property that $\{Tv'_\alpha\}$ converges. Thus, T takes bounded sets into compact sets, and we say that T is *compact*.

(b) If the sequence $\{v_\alpha\}$ is bounded and has the additional property that for each u in $V_{\mathscr{A}}$ the scalar product $\mathscr{A}(u, v_\alpha)$ converges to $\mathscr{A}(u, w)$, we say that v_α converges weakly to w. It then follows that Tv_α converges to Tw in norm, and we say that T is *completely continuous*.

These two concepts always coincide for a linear operator.

4. Inhomogeneous equations. We have reduced the eigenvalue problem to a problem of finding the successive maxima of the ratio of two quadratic forms. We wish to fit the inhomogeneous problem

$$Au = f \tag{4.1}$$

into the same framework. Here A is a linear transformation from V to W, f is a given element of W, and we wish to find u.

Suppose that we can find a sesquilinear functional (v, w) on $V \times W$ with the following properties.

HYPOTHESES 4.1. (a) $(v,w) = 0$ *for all* v *in* V *implies* $w = 0$.

(b) *The sesquilinear functional*

$$\mathscr{A}(u,v) \equiv (u, Av)$$

is Hermitian, and $\mathscr{A}(u,u)$ *is positive definite on* V.

(c) *The linear functional* (v,f) *on* V *is bounded in the sense that there is a constant* c *depending only on* f *so that*

$$|(v,f)| \leqq c\mathscr{A}(v,v)^{1/2} \quad \textit{for all } v \in V.$$

We know from the corollary to Theorem 4.2 of Chap. 2 that there exists a unique u in the completion $V_{\mathscr{A}}$ of the pre-Hilbert space V with scalar product $\mathscr{A}(u,v)$ with the property that the bounded linear functional (v,f) is represented in the form

$$\mathscr{A}(v,u) = (v,f). \tag{4.2}$$

Clearly, if u satisfies (4.1) it also satisfies (4.2). Conversely if u satisfies (4.2) and lies in V, then

$$(v, Au - f) = 0$$

for all v in V, and hence by Hypothesis 4.1(a), u satisfies (4.1). If the solution u of (4.2) does not lie in V, we call it a generalized solution $A^{-1}f$ of (4.1).

If A is an elliptic differential operator of the form (1.6) with smooth coefficients on a smooth domain D, if the assumptions of Example 1.5 hold, if f is a smooth function, and if

$$(u,v) = \int_D u\bar{v}\,dx,$$

the regularity theory for elliptic equations (see, e.g., Fichera (1965)) tells us that every generalized solution lies in V.

To construct a solution of (4.2), we imitate the proof of Theorem 4.2 of Chap. 2.

THEOREM 4.1. *Under the Hypotheses* 4.1, *the value*

$$\mu_1 = \sup_{\substack{v \in V \\ \mathscr{A}(v,v)=1}} |(v,f)|^2$$

is attained for some $v = u_1$ *in* $V_{\mathscr{A}}$. *The element*

$$u \equiv \overline{(u_1,f)}u_1$$

satisfies (4.2).

Proof. We define the sesquilinear functional

$$\mathscr{B}(u,v) = (u,f)\overline{(v,f)}.$$

Then the Hypotheses 1.1 are satisfied.

By Hypothesis 4.1 (a) above $f \neq 0$ implies $\mu_1 > 0$. The linear condition $(v,f) = 0$ makes $\mathscr{B}(v,v) = 0$. Hence by Theorem 2.5 there exists a u_1 in $V_{\mathscr{A}}$ with $\mathscr{A}(u_1,u_1) = 1, \mathscr{B}(u_1,u_1) = \mu_1$. By Theorem 2.1,

$$\mu_1\mathscr{A}(v,u_1) = \mathscr{B}(v,u_1) = (v,f)\overline{(u_1,f)}. \tag{4.3}$$

Setting $v = u_1$, we see that

$$\mu_1 = |(u_1, f)|^2.$$

Dividing (4.3) by $\overline{(u_1, f)}$ and recalling that $\mathscr{A}(v, u)$ is sesquilinear, we find that the function $u = \overline{(u_1, f)}u_1$ satisfies (4.2), which completes the proof.

We observe that μ_1 is the square of the norm of the linear functional (v, f).

Theorem 4.1 shows that if a sesquilinear functional (v, w) for which Hypotheses 4.1 are satisfied can be found, the solution of (4.1) is equivalent to finding the only nonzero eigenvalue and the corresponding eigenvector of a self-adjoint completely continuous operator T which is defined by

$$\mathscr{A}(v, Tw) = (v, f)\overline{(w, f)}$$

for all v in $V_{\mathscr{A}}$. Thus any method for approximating the eigenvalues and eigenvectors of such an operator also approximates the solution of the inhomogeneous problem (4.1).

We now note that if Hypotheses 1.1 hold, then every element of the form Au or Bu with u in V satisfies Hypothesis 4.1(c). In particular, then, the transformation $A^{-1}B$ from V to $V_{\mathscr{A}}$ is defined. Since $\mathscr{A}(v, A^{-1}Bu) = (v, Bu) = \mathscr{B}(v, u)$, we see that $A^{-1}Bu = Tu$ in V.

Since all elements of W of the form Au or Bu, with u in V satisfy Hypothesis 4.1(c), we may, without loss of generality, assume that all elements of W satisfy this hypothesis. (We simply eliminate those that don't.) Then A^{-1} is a linear transformation from W to $V_{\mathscr{A}}$.

We complete W to a Hilbert space with the norm $\|w\| = \mathscr{A}(A^{-1}w, A^{-1}w)^{1/2}$. Then for any u in V

$$\|Au\| = \mathscr{A}(u, u)^{1/2} = \|u\|,$$

and for any w in W,

$$\|A^{-1}w\| = \mathscr{A}(A^{-1}w, A^{-1}w)^{1/2} = \|w\|.$$

Thus the transformations A and A^{-1} are norm-preserving.

By continuity we extend A to a one-to-one norm-preserving transformation from $V_{\mathscr{A}}$ onto $W_{\mathscr{A}}$, and A^{-1} to the inverse transformation of A.

We also extend B to a bounded linear transformation from $V_{\mathscr{A}}$ into $W_{\mathscr{A}}$. Then $T = A^{-1}B$.

5. The Poincaré and Courant-Weyl principles. The characterization of the eigenvalues given by Theorem 2.3 is of limited computational use as it stands. This is because it is necessary to know the eigenvectors $u_1, \cdots, u_{n-1}$ exactly before μ_n can even be defined. We shall avoid this problem by means of either of two principles.

THEOREM 5.1 (Poincaré's principle (1890)). *Let the eigenvalues* $\mu_1 \geqq \mu_2 \geqq \cdots$ *be defined by* (2.4), *with the convention that if* μ_n *is not attained, we put* $\mu_n = \mu_{n+1} = \mu_{n+2} = \cdots$. *Then*

$$\mu_n = \sup_{\substack{v_1, \cdots, v_n \in V_{\mathscr{A}} \\ \text{linearly} \\ \text{independent}}} \left\{ \min_{c_1, \cdots, c_n} \frac{\mathscr{B}(\sum_1^n c_i v_i, \sum_1^n c_i v_i)}{\mathscr{A}(\sum_1^n c_i v_i, \sum_1^n c_i v_i)} \right\}. \tag{5.1}$$

Proof. Let $v_1, \cdots, v_n$ be any n linearly independent elements of $V_{\mathscr{A}}$. Then there exists at least one nontrivial linear combination $v = a_1 v_1 + \cdots + a_n v_n$ which satisfies the $n-1$ orthogonality conditions $\mathscr{A}(v, u_1) = \cdots = \mathscr{A}(v, u_{n-1}) = 0$. Hence by (2.4),

$$\frac{\mathscr{B}(\sum_1^n a_i v_i, \sum_1^n a_i v_i)}{\mathscr{A}(\sum_1^n a_i v_i, \sum_1^n a_i v_i)} \leqq \mu_n. \tag{5.2}$$

(If the sequence of eigenvalues $\mu_1 \geqq \mu_2 \geqq \cdots$ terminates at μ_k with $k < n$ so that the value $\mu_{k+1} = \cdots = \mu_n$ is not attained, we reach the same conclusion by making v orthogonal to $u_1, \cdots, u_k$.)

Since the v_i are linearly independent, the set of c_i for which $\mathscr{A}(\sum_1^n c_i v_i, \sum_1^n c_i v_i) = 1$ is closed and bounded. Hence $\mathscr{B}(\sum_1^n c_i v_i, \sum_1^n c_i v_i)$ takes on its minimum on this set. We see from (5.2) that

$$\min_{c_1, \cdots, c_n} \frac{\mathscr{B}(\sum_1^n c_i v_i, \sum_1^n c_i v_i)}{\mathscr{A}(\sum_1^n c_i v_i, \sum_1^n c_i v_i)} \leqq \mu_n \tag{5.3}$$

for any set of linear independent elements $v_i, \cdots, v_n$. If the eigenvalue μ_n is attained, the choice $v_i = u_i, i = 1, \cdots, n$, makes the left-hand side equal to μ_n, so that (5.1) is verified. If μ_n is not attained, we choose the v_i from the sequence with the properties (2.12) to obtain (5.1).

Remark. Poincaré's principle characterizes the number of positive (or nonnegative) eigenvalues of the Rayleigh quotient $\mathscr{B}(u, u)/\mathscr{A}(u, u)$ as the largest dimension (possibly zero or infinite) of a linear subspace of V on which $\mathscr{B}(u, u)$ is positive definite (or positive semidefinite). These numbers are clearly independent of the quadratic functional $\mathscr{A}(u, u)$. Similar reasoning shows that the multiplicity of the eigenvalue $\mu = 0$ is also independent of the functional $\mathscr{A}(u, u)$.

THEOREM 5.2 (Courant–Weyl principle) (Courant (1920), Weyl (1912)). *Let the sequence μ_n be defined by* (2.4) *with the convention that if the sequence terminates at μ_k we put $\mu_k = \mu_{k+1} = \cdots$. Then*

$$\mu_n = \min_{\substack{l_1, \cdots, l_{n-1} \\ \text{linear} \\ \text{functionals}}} \left\{ \sup_{\substack{v \in V \\ l_1(v) = l_2(v) = \cdots = l_{n-1}(v) = 0}} \frac{\mathscr{B}(v, v)}{\mathscr{A}(v, v)} \right\}. \tag{5.4}$$

Proof. Choose any linearly independent elements $v_1, \cdots, v_n$ in V. For a given set of linear functionals $l_1, \cdots, l_{n-1}$ there is a nontrivial linear combination $w = \sum_1^n a_i v_i$ which satisfies $l_1(w) = \cdots = l_{n-1}(w) = 0$. Hence,

$$\sup_{l_1(v) = \cdots = l_{n-1}(v) = 0} \frac{\mathscr{B}(v, v)}{\mathscr{A}(v, v)} \geqq \frac{\mathscr{B}(w, w)}{\mathscr{A}(w, w)} \geqq \min_{c_1 \cdots c_n} \frac{\mathscr{B}(\sum_1^n c_i v_i, \sum_1^n c_i v_i)}{\mathscr{A}(\sum_1^n c_i v_i, \sum_1^n c_i v_i)}$$

for any linearly independent set $v_1, \cdots, v_n$. By (5.1) we have

$$\sup_{l_1(v) = \cdots = l_{n-1}(v) = 0} \frac{\mathscr{B}(v, v)}{\mathscr{A}(v, v)} \geqq \mu_n. \tag{5.5}$$

This is true for any $l_1, \cdots, l_{n-1}$. If the eigenvectors $u_1, \cdots, u_{n-1}$ exist, we can choose $l_i(u) = \mathscr{A}(u, u_i)$, $i = 1, \cdots, n-1$, to make equality hold in (5.5), which then implies (5.4). If the eigenvector u_k, $k < n$, fails to exist, we make $l_i(u) = \mathscr{A}(u, u_i)$ for $i = 1, \cdots, k-1$ and choose any functionals $l_k, \cdots, l_{n-1}$ to reach the same conclusion.

Remark 1. It was pointed out by Peter Lax that Theorems 5.1 and 5.2 are equivalent. That is, one can finish the proof of the latter without resort to the definition (2.4) and one can prove Theorem 5.1 directly from Theorem 5.2.

Remark 2. Theorems 5.1 and 5.2 are often called the maximum–minimum and minimum–maximum principles. However, in problems where the quadratic functional $\mathscr{B}(u, u)$ is positive definite, one can define the eigenvalue $\lambda_n = 1/\mu_n$ of problem (1.1) by looking at successive minima of the Rayleigh quotient $\mathscr{A}(u, u)/\mathscr{B}(u, u)$. This process of inverting the Rayleigh quotient interchanges maxima and minima. In particular, it turns Theorem 5.1 into a minimum–maximum theorem and Theorem 5.2 into a maximum–minimum theorem, thereby causing great confusion.

6. A mapping principle. We shall establish a theorem which serves as the basis for several approximation methods.

Let V_1 and V_2 be Hilbert spaces with norms $\mathscr{A}_1(u, u)^{1/2}$ and $\mathscr{A}_2(p, p)^{1/2}$, respectively. We are given bounded quadratic functionals $\mathscr{B}_1(u, u)$ and $\mathscr{B}_2(p, p)$ on V_1 and V_2, respectively. We define $\mu_1^{(1)} \geqq \mu_2^{(2)} \geqq \cdots$ to be the eigenvalues of the Rayleigh quotient $\mathscr{B}_1(u, u)/\mathscr{A}_1(u, u)$ on V_1 and $\mu_1^{(2)} \geqq \mu_2^{(2)} \geqq \cdots$ to be the eigenvalues of $\mathscr{B}_2(p, p)/\mathscr{A}_2(p, p)$ on V_2. We shall prove an inequality between these eigenvalues under some conditions.

THEOREM 6.1 (Mapping principle). *Let M be a linear transformation from a subspace S_1 of V_1 into V_2. Suppose that for some nondecreasing functions $f(\xi)$ and $g(\xi)$ the inequalities*

$$\mathscr{B}_2(Mu, Mu) \geqq f\left(\frac{\mathscr{B}_1(u, u)}{\mathscr{A}_1(u, u)}\right)\mathscr{A}_2(Mu, Mu) \tag{6.1}$$

and

$$\mathscr{A}_2(Mu, Mu) \geqq g\left(\frac{\mathscr{B}_1(u, u)}{\mathscr{A}_1(u, u)}\right)\mathscr{A}_1(u, u) \tag{6.2}$$

hold for all nonzero u in S_1.

If S_1 contains the eigenfunctions $u_1^{(1)}, u_2^{(1)}, \cdots, u_n^{(1)}$ corresponding to $\mu_1^{(1)}, \mu_2^{(1)}, \cdots, \mu_n^{(1)}$ and if

$$g(\mu_n^{(1)}) > 0, \tag{6.3}$$

then

$$\mu_n^{(2)} \geqq f(\mu_n^{(1)}). \tag{6.4}$$

Proof. Let T_n be the set of linear combinations of $u_1^{(1)}, u_2^{(1)}, \cdots, u_n^{(1)}$, which is, by hypothesis, a subspace of S_1. Then

$$\mathscr{B}_1(u, u) \geqq \mu_n^{(1)} \mathscr{A}_1(u, u)$$

for u in T_n. Since $g(\mu_n^{(1)}) > 0$, (6.2) shows that for any nonzero u in T_n, $\mathscr{A}_2(Mu, Mu) > 0$ and hence $Mu \neq 0$. That is, the elements $Mu_1^{(1)}, \cdots, Mu_n^{(1)}$ are linearly independent.

Therefore by the Poincaré principle

$$\begin{aligned} \mu_n^{(2)} &\geqq \min_{u \in T_n} \frac{\mathscr{B}_2(Mu, Mu)}{\mathscr{A}_2(Mu, Mu)} \\ &\geqq \min_{u \in T_n} f\left(\frac{\mathscr{B}_1(u, u)}{\mathscr{A}_1(u, u)}\right) \\ &\geqq f(\mu_n^{(1)}). \end{aligned}$$

Remark 1. If $\mu_n^{(1)}$ is not attained, the result can still be proved in essentially the same way if S_1 contains a sequence of elements $v_1, v_2, \cdots$ of V_1 which satisfy (2.12) and if $g(\xi) > 0$ in a neighborhood of $\mu_n^{(1)}$.

Remark 2. The condition (6.2) was used only to show that if $u \neq 0$, then $Mu \neq 0$. It can therefore be replaced by the condition

$$\mathscr{B}_2(Mu, Mu) \geqq g\left(\frac{\mathscr{B}_1(u, u)}{\mathscr{A}_1(u, u)}\right) \mathscr{A}_1(u, u). \tag{6.5}$$

7. The first monotonicity principle and the Rayleigh–Ritz method. We let V_1 be a subspace of V_2, set $\mathscr{A}_1(u, u) = \mathscr{A}_2(u, u) = \mathscr{A}(u, u)$ and $\mathscr{B}_1(u, u) = \mathscr{B}_2(u, u) = \mathscr{B}(u, u)$, and let M be the injection mapping, which says to consider the element u of V_1 as an element of V_2. Then the hypotheses (6.1) and (6.2) are satisfied when $f(\xi) = \xi$, $g(\xi) \equiv 1$, and the mapping principle of the preceding section becomes the following theorem.

THEOREM 7.1 (First monotonicity principle). *Let V_2 be a Hilbert space with norm $\mathscr{A}(u, u)^{1/2}$ and let $\mathscr{B}(u, u)$ be a bounded quadratic functional on V_2. Let V_1 be a subspace of V_2. If $\mu_1^{(2)} \geqq \mu_2^{(2)} \geqq \cdots$ are the eigenvalues of the Rayleigh quotient $\mathscr{B}(u, u)/\mathscr{A}(u, u)$ on V_2 and $\mu_1^{(1)} \geqq \mu_2^{(2)} \geqq \ldots$ are the eigenvalues of the same Rayleigh quotient with u restricted to lie in V_1, then*

$$\mu_n^{(1)} \leqq \mu_n^{(2)}, \qquad n = 1, 2, \cdots. \tag{7.1}$$

An important application of the first monotonicity principle is the following theorem.

THEOREM 7.2 (Rayleigh–Ritz method). *Let $V_{\mathscr{A}}$ be the closure of V under the norm $\mathscr{A}(u, u)^{1/2}$, and let $\mu_1 \geqq \mu_2 \geqq \cdots$ be the eigenvalues of the Rayleigh quotient $\mathscr{B}(u, u)/\mathscr{A}(u, u)$ on $V_{\mathscr{A}}$. Choose k linearly independent elements $v_1, v_2, \cdots, v_k$ of $V_{\mathscr{A}}$. Let the eigenvalues of the matrix problem*

$$\sum_{j=1}^{k} \mathscr{B}(v_i, v_j)\bar{x}_j = \mu \sum_{j=1}^{k} \mathscr{A}(v_i, v_j)\bar{x}_j, \qquad i = 1, \cdots, k, \tag{7.2}$$

be $\mu'_1 \geqq \mu'_2 \geqq \cdots \geqq \mu'_k$. *Then*

$$\mu'_i \leqq \mu_i, \qquad i = 1, \cdots, k. \tag{7.3}$$

Proof. Let $V_2 = V_{\mathscr{A}}$, let V_1 be the space spanned by $v_1, \cdots, v_k$, and apply Theorem 7.1.

Remark. The process of approximating the eigenvalues μ_i by the lower bounds μ'_1 (or the eigenvalues $\lambda_i = 1/\mu_i$ by the upper bounds $\lambda'_i = 1/\mu'_i$) is known as the Rayleigh–Ritz method.[1]

The fact that the v_i need only be in the completion $V_{\mathscr{A}}$ rather than in V means that it is easier to find admissible functions.

Example 7.1. If we consider the problem

$$\begin{aligned} -\Delta u = \lambda u \quad & \text{in } D, \\ u = 0 \quad & \text{on } \Sigma_1, \\ \partial u/\partial u + \rho u = 0 \quad & \text{on } \Sigma_2 = \partial D - \Sigma_1, \quad \rho \geqq 0, \end{aligned}$$

and define $\mathscr{A}(u, v)$ and $\mathscr{B}(u, v)$ as in (2.20), we find that $V_{\mathscr{A}}$ is a subspace of the Sobolev space $H^1(D)$ which consists of functions which vanish on Σ_1 but need not satisfy the boundary condition on Σ_2. In particular, a function v_i which is continuous and vanishes on Σ_1 and which has square integrable piecewise continuous partial derivatives is admissible. The quadratic functional $\mathscr{A}(u, u)$ depends upon ρ.

In a problem like this we say that the condition $u = 0$ on Σ_1 is a *stable boundary condition*. The boundary condition on Σ_2, which disappears in $V_{\mathscr{A}}$ but which influences the quadratic functional $\mathscr{A}(u, u)$, is said to be an *unstable boundary condition* or a *natural boundary condition.*

In general, if we have a problem $Au = \lambda Bu$ where A is an elliptic operator of order $2M$ of the form (1.6), and if the assumptions in Example 1.5 are satisfied, then boundary conditions or linear combinations of boundary conditions which involve derivatives of order lower than M will be stable, while the rest are unstable.

For instance, in Chap. 1 in the problem (3.3) of the vibrating plate or the problem (3.8) of the buckling of a plate, boundary conditions of the form $u = 0$ and $\partial u/\partial n = 0$ are stable, while the no-force condition $P[u] = 0$ defined in (3.6), Chap. 1, and the no-torque condition $M[u] = 0$ defined in (3.5), Chap. 1, are

[1] Rayleigh, in his book *The Theory of Sound* (1878, §§ 88, 89), gave an argument to indicate that the Rayleigh quotient should give an upper bound for the lowest eigenvalue of a vibrating system with finitely many degrees of freedom. He assumed that the same was true for a continuous system, and computed several examples. In the second edition (1894, § 92a), Rayleigh stated the more general principle that any constraint raises all the eigenvalues. He gave a computational application in 1899.

Poincaré (1890, pp. 259–261) proved Theorem 7.2 as a computational device, but did not give any numerical examples.

Ritz (1909) found experimentally that the λ'_i gave remarkably good approximations for the eigenvalues and also for the eigenvectors of the vibrating string and the vibrating square plate with free edges. He does not seem to have been aware that his approximations to the higher eigenvalues were upper bounds.

unstable. Similarly, in the problem (3.2), Chap. 1, of a vibrating elastic medium, boundary conditions which involve only the displacements w_i are stable, while conditions which involve forces and hence derivatives of the w_i are unstable.

As another example of an application of Theorem 7.1 we consider the following.

Example 7.2. Take two domains D_1 and D_2 with $D_1 \subset D_2$, and consider the eigenvalue problems

$$-\Delta u = \lambda^{(2)} u \quad \text{in } D_2,$$

$$u = 0 \quad \text{on } \partial D_2,$$

and

$$-\Delta v = \lambda^{(1)} v \quad \text{in } D_1,$$

$$v = 0 \quad \text{on } \partial D_1.$$

We form the real bilinear forms

$$\mathscr{A}_2(u, v) = \int_{D_2} \operatorname{grad} u \cdot \operatorname{grad} v \, dx,$$

$$\mathscr{B}_2(u, v) = \int_{D_2} uv \, dx.$$

Completing the space $V = \{v \in C^\infty(\bar{D}_2) | v = 0 \text{ on } \partial D_2\}$, we obtain the subspace V_2 of $H^1(D_2)$ of functions which vanish on the boundary ∂D_2 of D_2.

We now let V_1 be the completion of the set of functions in V which vanish outside D_1. Then $V_1 \subset V_2$. The successive maxima $\mu_n^{(2)}$ and $\mu_n^{(1)}$ are the reciprocals of the eigenvalues $\lambda_n^{(2)}$ and $\lambda_n^{(1)}$ of the problems in D_2 and D_1. Hence, Theorem 7.1 shows that

$$\lambda_n^{(2)} \leqq \lambda_n^{(1)}, \qquad n = 1, 2, \cdots. \tag{7.4}$$

That is, shrinking the domain increases the eigenvalues. Such a result holds for a very general class of eigenvalue problems for differential operators with Dirichlet boundary conditions.

We note that it is important to deal with the boundary condition $u = 0$ on ∂D_1 or at least on the part of ∂D_1 which lies in D_2. Other conditions cannot be imposed by taking a subset. Therefore, the monotonicity (7.4) for λ_n as a functional of domain does not hold for other boundary conditions.

Example 7.3. Consider the two problems

$$-\Delta u + u = \lambda^{(2)} u \quad \text{in } D_2,$$

$$\partial u / \partial n = 0 \quad \text{on } \partial D_2,$$

$$-\Delta v + v = \lambda^{(1)} v \quad \text{in } D_1,$$

$$\partial v / \partial n = 0 \quad \text{on } \partial D_1,$$

where D_2 is the rectangle $-\pi < x_1 < \pi$, $-2\pi/3 < x_2 < 2\pi/3$, and D_1 is this rectangle with the square $|x_1| < \pi/3$, $0 < x_2 < 2\pi/3$ removed. We put

$$\mathscr{A}_1(u, u) = \int_{D_1} (|\text{grad } u|^2 + |u|^2)\, dx,$$

$$\mathscr{B}_1(u, u) = \int_{D_1} |u|^2\, dx.$$

It is easily seen that the lowest eigenvalues are both 1: $\lambda_1^{(1)} = \lambda_1^{(2)} = 1$ with the eigenfunctions $u_1 = v_1 = 1$. We can also see that $\lambda_2^{(2)} = \frac{5}{4}$ with $u_2^{(2)} = \sin \frac{1}{2}x_1$. A computation shows that $\mathscr{A}_1(u_2^{(2)}, v_1) = 0$ and that

$$\mathscr{A}_1(u_2^{(2)}, u_2^{(2)}) = \frac{25}{18}\pi^2 + \frac{\sqrt{3}}{4}\pi,$$

$$\mathscr{B}_1(u_2^{(2)}, u_2^{(2)}) = \frac{10}{9}\pi^2 + \frac{\pi}{\sqrt{3}}.$$

Hence

$$\mu_2^{(1)} \geqq \frac{\mathscr{B}_1(u_2^{(2)}, u_2^{(2)})}{\mathscr{A}_1(u_2^{(2)}, u_2^{(2)})} > \frac{4}{5} = \mu_2^{(2)}.$$

Consequently,

$$\lambda_2^{(1)} < \lambda_2^{(2)},$$

so that (7.4) does not hold.

Another application of Theorem 7.1 is the following.

Example 7.4. Let

$$\begin{aligned} -\Delta u &= \lambda^{(2)} u && \text{in } D, \\ u &= 0 && \text{on } \Sigma_1, \\ \partial u/\partial n + \rho u &= 0 && \text{on } \Sigma_2 = \partial D - \Sigma_1, \\ -\Delta v &= \lambda^{(1)} v && \text{in } D, \\ v &= 0 && \text{on } \partial D. \end{aligned}$$

Let

$$\mathscr{A}(u, v) = \int_D \text{grad } u \cdot \text{grad } \bar{v}\, dx + \int_{\Sigma_2} \rho u \bar{v}\, dS,$$

$$\mathscr{B}(u, v) = \int_D u\bar{v}\, dx.$$

The space $V_2 \equiv V_{\mathscr{A}}$ for the first problem is the subspace of the Sobolev space $H^1(D)$ of functions which vanish on Σ_1. Let V_1 be the subspace of $V_{\mathscr{A}}$ of functions which vanish on the whole boundary ∂D. The $1/\lambda_n^{(2)}$ are the successive maxima of the Rayleigh quotient on V_2. The $1/\lambda_n^{(1)}$ are the successive maxima on V_1. (Note that the integral over Σ_2 disappears for functions of V_1.) Thus by Theorem 7.1,

$$\lambda_n^{(1)} \geqq \lambda_n^{(2)}, \qquad n = 1, 2, \cdots.$$

A similar argument shows that, in general, adding more stable boundary conditions increases the λ_n.

8. The second monotonicity principle. We are given a linear vector space V together with two quadratic functionals $\mathscr{B}_1(u, u)$ and $\mathscr{B}_2(u, u)$ and two positive definite quadratic functionals $\mathscr{A}_1(u, u)$ and $\mathscr{A}_2(u, u)$, with the ratios $\mathscr{B}_1(u, u)/\mathscr{A}_1(u, u)$ and $\mathscr{B}_2(u, u)/\mathscr{A}_2(u, u)$ bounded. We let $\mu_1^{(\alpha)} \geqq \mu_2^{(\alpha)} \geqq \cdots$, $\alpha = 1, 2$, denote the eigenvalues of the Rayleigh quotient $\mathscr{B}_\alpha(u, u)/\mathscr{A}_\alpha(u, u)$. We observe that if

$$\begin{aligned} &\mathscr{B}_2(u, u) \geqq \mathscr{B}_1(u, u), \\ &\mathscr{A}_2(u, u) \leqq \mathscr{A}_1(u, u) \end{aligned} \tag{8.1}$$

then (6.1) and (6.5) are satisfied with $Mu \equiv u, f(\xi) = g(\xi) = \xi$, provided $\mu_n^{(1)} > 0$. Therefore the mapping principle shows that if $\mu_n^{(1)} > 0$, the inequality

$$\mu_n^{(1)} \leqq \mu_n^{(2)} \tag{8.2}$$

holds. If $\mu_n^{(1)} \leqq 0$ but $\mu_n^{(2)} \geqq 0$, the inequality (8.2) is clearly satisfied. Thus we have the following theorem.

THEOREM 8.1. *Let the inequalities* (8.1) *be satisfied for all u in V. Then*

$$\mu_n^{(1)} \leqq \mu_n^{(2)}$$

whenever $\mu_n^{(2)} \geqq 0$.

This theorem states that increasing the numerator or decreasing the denominator of the Rayleigh quotient tends to increase the positive eigenvalues.

Remark. If $\mathscr{A}_2(u, u) = \mathscr{A}_1(u, u)$ for all u, then (6.1) and (6.2) with $f(\xi) = \xi, g(\xi) \equiv 1$ are satisfied, and hence (8.2) holds even when $\mu_n^{(2)}$ is negative.

Example 8.1. Consider the problem

$$\begin{aligned} -\Delta u &= \lambda u \quad \text{in } D, \\ \partial u/\partial n + ku &= 0 \quad \text{on } \partial D, \quad k \geqq 0 \text{ constant}. \end{aligned} \tag{8.3}$$

Let

$$\mathscr{A}_1(u, u) = \int_D |\text{grad } u|^2 \, dx + k \int_{\partial D} |u|^2 \, dS,$$

$$\mathscr{B}_1(u, u) = \int_D |u|^2 \, dx.$$

Suppose we replace k by a smaller constant $\tilde{k}$ and call the resulting quadratic functional $\mathscr{A}_2(u, u)$. We let $\mathscr{B}_2(u, u) = \mathscr{B}_1(u, u)$. Then the inequalities (8.1) hold, and hence $\mu_n^{(2)} \geqq \mu_n^{(1)}$. In terms of the eigenvalues $\lambda_n(k) = 1/\mu_n$, we can say that *each* $\lambda_n(k)$ *is a nondecreasing function of* k.

On the other hand, it follows from the first monotonicity principle that each eigenvalue $\lambda_n(k)$ is bounded above by the corresponding eigenvalue λ'_n of the problem with boundary condition $u = 0$.

9. A separation theorem and an inequality for the eigenvalues of a sum of operators. We present a theorem which limits the amount by which the monotonicity principles can shift the eigenvalues.

THEOREM 9.1 (Separation theorem). *Let the quadratic forms $\mathscr{B}_1(u, u)$ and $\mathscr{B}_2(u, u)$ and the positive definite quadratic forms $\mathscr{A}_1(u, u)$ and $\mathscr{A}_2(u, u)$ be defined on V_1 and V_2 respectively. Let the Rayleigh quotients $\mathscr{B}_\alpha(u, u)/\mathscr{A}_\alpha(u, u)$ be bounded, and denote their eigenvalues by $\mu_1^{(\alpha)} \geqq \mu_2^{(\alpha)} \geqq \cdots$. If there are linear functionals $l_1, \cdots, l_r$ on V_2 so that*

$$v \in V_2, \qquad l_1(v) = l_2(v) = \cdots = l_r(v) = 0$$

implies

$$v \in V_1$$

and

$$\frac{\mathscr{B}_1(v, v)}{\mathscr{A}_1(v, v)} = \frac{\mathscr{B}_2(v, v)}{\mathscr{A}_2(v, v)},$$

then

$$\mu_{n+r}^{(2)} \leqq \mu_n^{(1)}, \qquad n = 1, 2, \cdots. \tag{9.1}$$

Proof. Let $u_1^{(1)}, u_2^{(1)}, \cdots$ be the eigenvectors corresponding to $\mu_1^{(1)} \geqq \mu_2^{(1)} \geqq \cdots$. By the Courant–Weyl principle

$$\begin{aligned} \mu_{n+r}^{(2)} &\leqq \sup_{\substack{v \in V_2 \\ l_1(v) = \cdots = l_r(v) = 0 \\ \mathscr{A}_1(v, u_1^{(1)}) = \cdots = \mathscr{A}_1(v, u_{n-1}^{(1)}) = 0}} \frac{\mathscr{B}_2(v, v)}{\mathscr{A}_2(v, v)} \\ &\leqq \sup_{\substack{v \in V_1 \\ \mathscr{A}_1(v, u_1^{(1)}) = \cdots = \mathscr{A}_1(v, u_{n-1}^{(1)}) = 0}} \frac{\mathscr{B}_1(v, v)}{\mathscr{A}_1(v, v)} \\ &= \mu_n^{(1)}. \end{aligned} \tag{9.2}$$

Remark 1. If $\mu_n^{(1)}$ is not attained, one finds the same result by using as many eigenfunctions $u_i^{(1)}$ as there are.

Remark 2. The idea of this theorem can be found in the works of Rayleigh (1877, 2nd ed. of 1894, § 92a) and Weyl (1912).

We note two particular cases of this theorem.

COROLLARY 1. *If $\mathscr{A}_1 = \mathscr{A}_2$ and $\mathscr{B}_1 = \mathscr{B}_2$ and if*

$$V_1 = \{v \in V_2 | l_1(v) = \cdots = l_r(v) = 0\},$$

then

$$\mu_{n+r}^{(2)} \leqq \mu_n^{(1)} \leqq \mu_n^{(2)}. \tag{9.3}$$

Proof. This is simply a combination of Theorem 9.1 and the first monotonicity principle.

COROLLARY 2. *If* $V_1 = V_2$, $\mathscr{A}_2(u, u) \leqq \mathscr{A}_1(u, u)$, *and* $\mathscr{B}_2(u, u) \geqq \mathscr{B}_1(u, u) \geqq 0$, *and if*

$$l_1(v) = \cdots = l_r(v) = 0$$

implies that

$$\mathscr{B}_1(v, v) = \mathscr{B}_2(v, v), \qquad \mathscr{A}_1(v, v) = \mathscr{A}_2(v, v),$$

then

$$\mu_{n+r}^{(2)} \leqq \mu_n^{(1)} \leqq \mu_n^{(2)}. \tag{9.4}$$

Proof. This is a combination of Theorem 9.1 and the second monotonicity principle.

Example 9.1. Consider the differential equation

$$-\frac{d}{dx}\left[(1 + x^2)\frac{du}{dx}\right] + xu = \lambda^{(1)}u \quad \text{for } 0 < x < 1,$$

$$u(0) = 0, \qquad u'(1) + 2u(1) = 0$$

and

$$-\frac{d}{dx}\left[(1 + x^2)\frac{du}{dx}\right] + xu = \lambda^{(2)}u \quad \text{for } 0 < x < 1,$$

$$u'(0) = 0, \qquad u'(1) + u(1) = 0.$$

We set

$$\mathscr{A}_1(u, u) = \int_0^1 [(1 + x^2)u'^2 + xu^2]\, dx + 4u(1)^2,$$

$$\mathscr{B}_1(u, u) = \int_0^1 u^2\, dx,$$

$$\mathscr{A}_2(u, u) = \int_0^1 [(1 + x^2)u'^2 + xu^2]\, dx + 2u(1)^2,$$

$$\mathscr{B}_2(u, u) = \int_0^1 u^2\, dx.$$

The space V_1 consists of functions in $H^1([0, 1])$ which vanish at $x = 0$. V_2 is just $H^1([0, 1])$. Thus $V_1 \subset V_2$ and $\mathscr{A}_1(u, u) \geqq \mathscr{A}_2(u, u)$, while $\mathscr{B}_1(u, u) = \mathscr{B}_2(u, u)$. The first and second monotonicity principles show that $\lambda_n^{(1)} \geqq \lambda_n^{(2)}$.

On the other hand, we note that when $u(0) = u(1) = 0$, $u \in V_1$ and $\mathscr{A}_1(u, u) = \mathscr{A}_2(u, u)$. Hence by Theorem 9.1 applied to $\mu_n^{(\alpha)} = 1/\lambda_n^{(\alpha)}$ we see that

$$\lambda_n^{(2)} \leqq \lambda_n^{(1)} \leqq \lambda_{n+2}^{(2)}$$

In this way one can find various separation theorems for Sturm–Liouville problems, some of which are classical. The idea of obtaining separation theorems in this way is due to A. Weinstein (1951), (1963). See also Weinberger (1955).

We now prove a useful inequality for the eigenvalues of a sum of operators.

THEOREM 9.2 (Weyl (1912)). *Let*

$$\mathscr{B}(u,u) = \mathscr{B}_1(u,u) + \mathscr{B}_2(u,u).$$

Denote by $\mu_1 \geqq \mu_2 \geqq \cdots$, $\mu_1^{(1)} \geqq \mu_2^{(1)} \geqq \cdots$, *and* $\mu_1^{(2)} \geqq \mu_2^{(2)} \geqq \cdots$ *the eigenvalues of the Rayleigh quotients* $\mathscr{B}(u,u)/\mathscr{A}(u,u)$, $\mathscr{B}_1(u,u)/\mathscr{A}(u,u)$, *and* $\mathscr{B}_2(u,u)/\mathscr{A}(u,u)$, *respectively. Then*

$$\mu_{k+l-1} \leqq \mu_k^{(1)} + \mu_l^{(2)}, \qquad k, l = 1, 2, \cdots. \tag{9.5}$$

Proof. By the Courant–Weyl principle

$$\begin{aligned} \mu_{k+l-1} &\leqq \sup_{\substack{\mathscr{A}(v,u_1^{(1)}) = \cdots = \mathscr{A}(v,u_{k-1}^{(1)}) = 0 \\ \mathscr{A}(v,u_1^{(2)}) = \cdots = \mathscr{A}(v,u_{l-1}^{(2)}) = 0}} \frac{\mathscr{B}(v,v)}{\mathscr{A}(v,v)} \\ &\leqq \sup \frac{\mathscr{B}_1(v,v)}{\mathscr{A}(v,v)} + \sup \frac{\mathscr{B}_2(v,v)}{\mathscr{A}(v,v)} \\ &\leqq \mu_k^{(1)} + \mu_l^{(2)}. \end{aligned}$$

Remark 1. If $\mu_k^{(1)}$ or $\mu_l^{(2)}$ is not attained, one finds the same result by using as many eigenfunctions as there are.

Remark 2. If $\mathscr{B}(u,u)$ is positive definite and $\mathscr{A}(u,u) = \mathscr{A}_1(u,u) + \mathscr{A}_2(u,u)$, we can obtain the analogous result

$$\lambda_{k+l-1} \geqq \lambda_k^{(1)} + \lambda_l^{(2)} \tag{9.6}$$

for the eigenvalues in ascending order of the Rayleigh quotients $\mathscr{A}/\mathscr{B}$, $\mathscr{A}_1/\mathscr{B}$, and $\mathscr{A}_2/\mathscr{B}$.

CHAPTER 4

Improvable Bounds for Eigenvalues

1. A priori and a posteriori error bounds. We have seen that lower bounds for the successive maxima $\mu_1 \geqq \mu_2 \geqq \cdots$ of the quotient $\mathscr{B}(u, u)/\mathscr{A}(u, u)$ can be obtained by the Rayleigh–Ritz method. One must only choose elements $v_1, \cdots, v_k$ in $V_{\mathscr{A}}$ and find the eigenvalues of the symmetric matrix problem (7.2), Chap. 3. There are, of course, computational difficulties associated with finding the eigenvalues of a large matrix, but we postpone a discussion of these to the chapter on finite difference equations.

A serious problem arises from the fact that one knows that the matrix eigenvalues are lower bounds, but not how close these are to the actual eigenvalues. There are two ways to remedy this situation:

(a) Find a bound for the difference $\mu_n - \mu_n'$ between the correct eigenvalue μ_n and the Rayleigh–Ritz approximation μ_n'.

(b) Find an upper bound $\tilde{\mu}_n$ for μ_n.

These two ideas are, of course, equivalent in the sense that if $\tilde{\mu}_n$ is known, $\tilde{\mu}_n - \mu_n'$ is a bound for the error $\mu_n - \mu_n'$ while if an error bound $\mu_n - \mu_n' \leqq \varepsilon_n$ is known, then $\mu_n' + \varepsilon_n$ is an upper bound for μ_n.

Computationally, there are two more important distinctions:

(a) A method may give one bound with no idea about how to obtain a closer bound, or it may give an improvable bound in the sense that there is an algorithm to improve the result if one is willing to exert the effort to do so. We shall treat the second kind of bound in this chapter. We shall discuss some methods of the first kind in the last chapter.

(b) An error bound may be a priori or a posteriori. If we have two improvable methods, one giving upper bounds and one lower bounds, and a convergence proof, we know that with enough effort the difference between the upper and lower bounds can be reduced to an arbitrarily small amount. However, we have no idea of how much effort is required for this accuracy, and the computation involves trial and error. Many unsatisfactory bounds may have to be computed to arrive at a satisfactory one.

On the other hand, an a priori bound tells us before we start the computation how much effort will be required. It is not always possible to do this. and we shall give improvable bounds of both kinds in the following sections.

2. Error bounds for eigenvalues in the Rayleigh–Ritz method. We wish to bound the error $\mu_n - \mu_n'$ where μ_n is the nth eigenvalue of the Rayleigh quotient $\mathscr{B}(u, u)/\mathscr{A}(u, u)$ and μ_n' is the nth eigenvalue of the corresponding problem with $V_{\mathscr{A}}$ replaced by the set of linear combinations of the chosen elements $v_1, v_2, \cdots, v_k$.

We shall assume that we know that for some $n \leqq k$ the eigenvectors $u_1, \cdots, u_n$ exist and that to each of these eigenvectors u_i there corresponds a linear combination w_i of $v_1, \cdots, v_k$ which approximates u_i in the sense that

$$\frac{\mathscr{A}(u_i - w_i, u_i - w_i)}{\mathscr{A}(u_i, u_i)} \leqq \eta_i^2. \tag{2.1}$$

Moreover, we suppose that the η_i are so small that

$$\sum_{i=1}^{n} \eta_i^2 < 1. \tag{2.2}$$

We shall prove the following error bound.

THEOREM 2.1. *Let $\mu_1 \geqq \mu_2 \geqq \cdots$ be the eigenvalues of the Rayleigh quotient $\mathscr{B}(u, u)/\mathscr{A}(u, u)$ on $V_{\mathscr{A}}$. Let $\mu_1' \geqq \mu_2' \geqq \cdots \geqq \mu_k'$ be the eigenvalues on the subspace of $V_{\mathscr{A}}$ spanned by $v_1, v_2, \cdots, v_k$. Let $-\gamma$ be a number so that $\mathscr{B}(u, u) \geqq -\gamma\mathscr{A}(u, u)$ for all u in V.*

If there are linear combinations $w_1, w_2, \cdots, w_n$ of the v_i which approximate the eigenvectors $u_1, \cdots, u_n$ in the sense of (2.1) and if the bounds η_i are so small that (2.2) holds, then

$$0 \leqq \mu_n - \mu_n' \leqq (\mu_n + \gamma) \sum_{i=1}^{n} \eta_i^2. \tag{2.3}$$

Proof. The first inequality follows from the first monotonicity principle.

In order to obtain the second inequality, we wish to apply the mapping principle of § 6, Chap. 3. We define the mapping

$$Mu_i = w_i.$$

Let

$$u = \sum_{i=1}^{n} d_i u_i.$$

By the triangle inequality and Schwarz's inequality we see that

$$\begin{aligned}
\mathscr{A}(Mu - u, Mu - u) &= \mathscr{A}\left(\sum_{i=1}^{n} d_i(w_i - u_i), \sum_{i=1}^{n} d_i(w_i - u_i)\right) \\
&\leqq \left\{\sum_{i=1}^{n} |d_i|(\mathscr{A}(w_i - u_i, w_i - u_i))^{1/2}\right\}^2 \\
&\leqq \sum_{i=1}^{n} |d_i|^2 \mathscr{A}(u_i, u_i) \sum_{i=1}^{n} \frac{\mathscr{A}(w_i - u_i, w_i - u_i)}{\mathscr{A}(u_i, u_i)} \\
&\leqq \mathscr{A}(u, u) \sum_{i=1}^{n} \eta_i^2.
\end{aligned} \tag{2.4}$$

We then see from the triangle inequality that

$$\mathscr{A}(u,u)^{1/2} \leqq \mathscr{A}(Mu,Mu)^{1/2} + \mathscr{A}(Mu-u,Mu-u)^{1/2}$$
$$\leqq \mathscr{A}(Mu,Mu)^{1/2} + \left(\sum_{i=1}^{n} \eta_i^2\right)^{1/2} \mathscr{A}(u,u)^{1/2}.$$

Because of (2.2) this gives

$$\mathscr{A}(Mu,Mu) \geqq \left\{1 - \left(\sum_{i=1}^{n} \eta_i^2\right)^{1/2}\right\}^2 \mathscr{A}(u,u).$$

Thus we have (6.2), Chap. 3, with $g(\xi) \equiv \{1 - (\sum \eta_i^2)^{1/2}\}^2$.

We now observe that if for any v in V we define

$$P_n v = \sum_{i=1}^{n} \frac{\mathscr{A}(v,u_i)}{\mathscr{A}(u_i,u_i)} u_i$$

to be the orthogonal projection of v onto the span of $u_1, \cdots, u_n$, then

$$(2.5) \qquad \begin{aligned} \mathscr{A}(v,v) &= \mathscr{A}(P_n v, P_n v) + \mathscr{A}(v - P_n v, v - P_n v), \\ \mathscr{B}(v,v) &= \mathscr{B}(P_n v, P_n v) + \mathscr{B}(v - P_n v, v - P_n v). \end{aligned}$$

Moreover, since $P_n v$ is a linear combination of $u_1, \cdots, u_n$ we see that

$$(2.6) \qquad \begin{aligned} \mathscr{B}(P_n v, P_n v) &\geqq \mu_n \mathscr{A}(P_n v, P_n v) \\ &= \mu_n \mathscr{A}(v,v) - \mu_n \mathscr{A}(v - P_n v, v - P_n v). \end{aligned}$$

If we apply (2.5) and (2.6) to the element $v = Mu$, we see that

$$\begin{aligned} \mathscr{B}(Mu,Mu) &\geqq \mu_n \mathscr{A}(Mu,Mu) - \mu_n \mathscr{A}(Mu - P_n Mu, Mu - P_n Mu) \\ &\quad + \mathscr{B}(Mu - P_n Mu, Mu - P_n Mu) \\ &\geqq \mu_n \mathscr{A}(Mu,Mu) - (\mu_n + \gamma) \mathscr{A}(Mu - P_n Mu, Mu - P_n Mu). \end{aligned}$$

Thus,

$$(2.7) \qquad \frac{\mathscr{B}(Mu,Mu)}{\mathscr{A}(Mu,Mu)} \geqq \mu_n - (\mu_n + \gamma) \frac{\mathscr{A}(Mu - P_n Mu, Mu - P_n Mu)}{\mathscr{A}(Mu,Mu)}.$$

In order to establish (6.1), Chap. 3, we must bound the ratio on the right from above. Let us define

$$(2.8) \qquad \alpha^2 = \frac{\mathscr{A}(Mu - P_n Mu, Mu - P_n Mu)}{\mathscr{A}(u,u)}.$$

We recall that $P_n u = u$ so that $u = P_n Mu - P_n(Mu - u)$. We then see from (2.5) and the triangle inequality that

$$(2.9) \qquad \begin{aligned} \mathscr{A}(Mu,Mu) &= \mathscr{A}(P_n Mu, P_n Mu) + \alpha^2 \mathscr{A}(u,u) \\ &\geqq [\mathscr{A}(u,u)^{1/2} - \mathscr{A}(P_n(Mu-u), P_n(Mu-u))^{1/2}]^2 + \alpha^2 \mathscr{A}(u,u). \end{aligned}$$

Using (2.4), (2.5), and (2.8), we have

$$\mathcal{A}(P_n(Mu-u), P_n(Mu-u)) = \mathcal{A}(Mu-u, Mu-u) - \mathcal{A}(Mu-P_nMu, Mu-P_nMu) \leqq \left(\sum_{i=1}^{n} \eta_i^2 - \alpha^2\right)\mathcal{A}(u,u).$$

Therefore from (2.9),

$$\mathcal{A}(Mu, Mu) \geqq \mathcal{A}(u,u)\left\{\left[1 - \left(\sum_{i=1}^{n} \eta_i^2 - \alpha^2\right)^{1/2}\right]^2 + \alpha^2\right\}.$$

The definition (2.8) of α^2 now shows that

$$\frac{\mathcal{A}(Mu - P_nMu, Mu - P_nMu)}{\mathcal{A}(Mu, Mu)} \leqq \frac{\alpha^2}{[1 - (\sum_{i=1}^{n} \eta_i^2 - \alpha^2)^{1/2}]^2 + \alpha^2}.$$

We maximize the right-hand side with respect to $0 \leqq \alpha^2 \leqq \sum_{i=1}^{n} \eta_i^2$. The maximum is just $\sum_{i=1}^{n} \eta_i^2$. Thus we find from (2.7) that

$$\frac{\mathcal{B}(Mu, Mu)}{\mathcal{A}(Mu, Mu)} \geqq \mu_n - (\mu_n + \gamma) \sum_{i=1}^{n} \eta_i^2.$$

This is (6.1), Chap. 3, with $f(\xi) \equiv \mu_n - (\mu_n + \gamma)\sum_{i=1}^{n} \eta_i^2$. The mapping principle now gives the second inequality in (2.3), and the theorem is proved.

Remark 1. If $\mathcal{B}(u, u)$ is positive definite, we may take $\gamma = 0$. If we use the eigenvalues $\lambda_n = 1/\mu_n$ and $\lambda'_n = 1/\mu'_n$, (2.3) becomes

$$(2.10) \qquad (1 - \sum \eta_i^2)\lambda'_n \leqq \lambda_n \leqq \lambda'_n.$$

This inequality is true but trivial when (2.2) is violated.

Remark 2. By a proof along similar lines Birkhoff, deBoor, Swartz and Wendroff (1966) obtained the error estimate

$$(2.11) \qquad \lambda_n \leqq \lambda'_n \leqq \lambda_n + \frac{\sum_{i=1}^{n} \lambda_i \eta_i^2}{\{1 - (\sum_{i=1}^{n} \mathcal{B}(u_i - w_i, u_i - w_i)/\mathcal{B}(u_i, u_i))^{1/2}\}^2}.$$

Whether this bound is better or worse than (2.10) depends upon the available error estimates. For the inequality (2.11) one must assume that the sum that occurs in the denominator is less than one, but not (2.2). The bound (2.11) requires bounds for $\mathcal{B}(u_i - w_i, u_i - w_i)$ as well as $\mathcal{A}(u_i - w_i, u_i - w_i)$.

Remark 3. While the need for an approximability estimate was implicit in the convergence proof for the Rayleigh–Ritz method and in the error bounding techniques of Krylov, this need was explicitly pointed out and emphasized by Birkhoff, deBoor, Swartz and Wendroff (1966).

Remark 4. We note that the error $\mu_i - \mu'_i$ in the eigenvalue is proportional to a sum of squares of norms of the errors $u_i - w_i$ in the eigenvectors. This means that fairly crude guesses about the eigenvectors lead to good approximations for the eigenvalues. This fact accounts for much of the success of the Rayleigh–Ritz method.

The upper bound (2.3) can be written as

$$\mu_n \leqq \frac{\mu_n' + \gamma \sum_{i=1}^n \eta_i^2}{1 - \sum_{i=1}^n \eta_i^2}.$$

A suitable constant γ can usually be found. The problem of bounding the error $\mu_n - \mu_n'$ is thus reduced to one of obtaining the bounds (2.1). That is, we need to know how well each eigenfunction u_i can be approximated by linear combinations of the functions $v_1, \cdots, v_k$ that have been chosen.

A simple consequence is the following: Suppose $\{v_j\}$ is a complete sequence in the sense that any element of V can be approximated with arbitrary accuracy in the norm $\mathscr{A}(u, u)^{1/2}$ by some finite linear combination of the v_j. Then for any $\varepsilon > 0$ there is an integer k so that if $\mu_n^{(k)}$ is the Rayleigh–Ritz bound for μ_n that comes from using the elements $v_1, \cdots, v_k$, then $\mu_n - \mu_n^{(k)} < \varepsilon$. That is, *for each n, $\mu_n^{(k)}$ converges to μ_n*. However, this statement does not tell us how k depends on ε.

A better result can often be obtained if one can find an a priori bound for a norm of some derivatives of the eigenfunctions u_i. We illustrate this idea with an example of the type treated by Krylov (1931).

Example 2.1. Consider the eigenvalue problem

$$\begin{aligned} -d^2u/dx^2 &= \lambda\rho(x)u \quad \text{in } [0, 1], \\ u(0) &= u(1) = 0. \end{aligned} \tag{2.12}$$

We suppose that $\rho(x) \geqq 0$ and take

$$\mathscr{A}(u, u) = \int_0^1 |u'|^2\, dx, \qquad \mathscr{B}(u, u) = \int_0^1 \rho|u|^2\, dx.$$

We observe that for any solution u of (2.12),

$$\begin{aligned} \int_0^1 |u''|^2\, dx = \lambda^2 \int \rho^2 |u|^2\, dx &\leqq \lambda^2 \max|\rho(x)|\mathscr{B}(u, u) \\ &= |\lambda| \max|\rho(x)|\mathscr{A}(u, u). \end{aligned}$$

Thus, if u_n is the eigenfunction corresponding to $\lambda_n = 1/\mu_n$, with $\mathscr{A}(u_n, u_n) = 1$, we have the inequality

$$\int |u_n''|^2\, dx \leqq |\lambda_n| \max|\rho(x)|.$$

We choose $v_j = \sin j\pi x, j = 1, \cdots, k$. Let w_j be the kth partial sum of the Fourier sine series for u_j,

$$w_j = \sum_{l=1}^k c_{lj} \sin l\pi x,$$

where

$$c_{lj} = 2\int_0^1 u_j \sin l\pi x\,dx.$$

Parseval's equation for this Fourier series gives

$$\mathscr{A}(u_j - w_j, u_j - w_j) = \frac{1}{2}\sum_{l=k+1}^{\infty} l^2\pi^2|c_{lj}|^2,$$

and

$$\int |u_j''|^2\,dx = \frac{1}{2}\sum_{l=1}^{\infty} l^4\pi^4|c_{lj}|^2.$$

Combining these two, we see that

$$\begin{aligned}\mathscr{A}(u_j - w_j, u_j - w_j) &\leqq \frac{1}{\pi^2(k+1)^2}\int |u_j''|^2\,dx\\ &\leqq \frac{|\lambda_j|\max|\rho|}{\pi^2(k+1)^2}.\end{aligned}$$

We thus have the inequality (2.1) with

$$\eta_j^2 = \frac{|\lambda_j|\max|\rho|}{\pi^2(k+1)^2}.$$

For any particular n we can make $\sum_1^n \eta_j^2$ arbitrarily small by choosing k sufficiently large. We note that since $\rho(x) \geqq 0$, we can take $\gamma = 0$ and the product $\mu_n \sum_i^n \lambda_j$ is bounded by n. Thus, (2.3) gives the bound

$$\mu_n' \leqq \mu_n \leqq \mu_n' + \frac{n\max\rho(x)}{\pi^2(k+1)^2}. \tag{2.13}$$

We observe that by (2.12), u_j'' vanishes at both endpoints. It has the Fourier sine series

$$u_j'' = -\sum l^2\pi^2 c_{lj}\sin l\pi x.$$

If ρ is continuously differentiable and uniformly positive, we then see that

$$\begin{aligned}\int_0^1 u_j'''^2\,dx &= \lambda_j^2\int_0^1 |\rho u' + \rho' u|^2\,dx\\ &\leqq 2\lambda_j^2\left\{\max \rho^2\mathscr{A}(u_j, u_j) + \max\frac{\rho'^2}{\rho}\mathscr{B}(u_j, u_j)\right\},\\ &= 2\lambda_j\left\{\lambda_j\max\rho^2 + \max\frac{\rho'^2}{\rho}\right\}.\end{aligned}$$

Now by Parseval's equation

$$\int_0^1 |u'''|^2\,dx = \frac{1}{2}\sum_1^\infty l^6\pi^6|c_{lj}|^2,$$

so that

$$\mathscr{A}(u_j - w_j, u_j - w_j) \leqq \frac{2\lambda_j}{\pi^4(k+1)^4}\left\{\lambda_j \max\rho^2 + \max\frac{\rho'^2}{\rho}\right\} \equiv \eta_j^2.$$

This η_j^2 is smaller than the previous one when k is sufficiently large. Whether it is better for a particular value of k that is considered computationally practical depends, of course, on the sizes of ρ and ρ'^2/ρ. Since λ_j occurs quadratically in η_j^2, we cannot eliminate it. Replacing all λ_j by λ_n and putting $\gamma = 0$, we obtain the bound

$$\mu_n \leqq \mu_n' + \frac{2n}{\pi^4(k+1)^4}\left\{\frac{\max\rho^2}{\mu_n} + \max\frac{\rho'^2}{\rho}\right\}.$$

This gives a quadratic inequality which may be solved for μ_n.

The method used to show that the sine functions give good approximations to the eigenfunctions of (2.12) may be extended in two ways:

(a) If the boundary conditions in (2.8) are changed to, say, $u(0) = 0, u'(1) + 2u(1) = 0$, we can no longer obtain Parseval's equation for $\int u''^2\,dx$ in terms of the sine series or even the Fourier series for $u(x)$. However, if we extend u_j by defining

$$u_j(x) = (1 - \alpha^{-2}(1-x)^2)[3u_j(2-x) - 2u_j(3-2x)] \quad \text{for } 1 < x < 1 + \alpha,$$

where $0 < \alpha \leqq \frac{1}{2}$, we find that $u_j(x)$ is continuously differentiable and has piecewise continuous second derivatives on the interval $(0, 1+\alpha)$, and that it vanishes at the endpoints. Moreover, one can bound the integral of $|u''|^2$ over the larger interval in terms of $\mathscr{A}(u_j, u_j)$ and $\mathscr{B}(u_j, u_j)$ for the unextended eigenfunction u_j. We now use the functions $v_j = \sin j\pi x/(1+\alpha)$, $j = 1, \cdots, k$, to obtain a bound for the integral of $|u_j - w_j|^2$ in terms of $\int |u''|^2\,dx$ just as before. Again the bounds η_j^2 are proportional to $(k+1)^{-2}$ and hence can be made arbitrarily small.

The same ideas can be carried over to partial differential equations. The extension across the boundary is still possible if the boundary is smooth, although the computations may become quite cumbersome. An added difficulty lies in the fact that there are many partial derivatives of a particular order and the differential equation gives only one linear combination of them. However, if A is of the form (1.6), Chap. 3, if the assumptions of Example 1.5 of Chap. 3 are satisfied, if B is of lower order, and if the coefficients and boundary are smooth, then the Sobolev norm $\|u_j\|_{H^s}$ for any s can be explicitly bounded. (See, e.g., the book of Fichera (1965).) If we extend the function across the boundary in such a way that it is defined everywhere and vanishes outside a bounded set, then we obtain explicit expansion theorems and Parseval equations in terms of products of sines or many other systems of functions. Such ideas will be discussed in § 2, Chap. 6, in connection with the finite element method.

(b) Another natural generalization of the Krylov bounds is the following. We observe that the functions $\sin j\pi x$ are the eigenfunctions of the problem

$$-u'' = \lambda u$$

with the same boundary conditions $u(0) = u(1) = 0$. More generally, we might have a given problem

$$Au = \lambda u$$

in a space V and a related problem

$$A_0 v = \lambda v$$

whose eigenvalues and eigenvectors can be found. If $A - A_0$ is in some sense smaller than A, the eigenvectors $v_1, v_2, \cdots, v_k$ of the second problem will yield bounds of the form (2.2) which can be made arbitrarily small by making k large. We shall discuss this idea in § 10.

3. The Weinstein–Aronszajn theorems on finite-dimensional perturbation. In this section we present some theorems which constitute the heart of Weinstein and Aronszajn methods of intermediate problems. The basic idea is due to A. Weinstein (1935), (1937).

We present the results in a somewhat more general setting. Let A_0 and B_0 be linear transformations from a linear vector space V to a space W.

We consider a complex number μ, and we suppose that the transformation $B_0 - \mu A_0$ is a one-to-one mapping from V to W; that is, for each w in W there is a unique v in V such that

$$B_0 v - \mu A_0 v = w. \tag{3.1}$$

We then say that μ is in the *resolvent set*, and we define the *resolvent* $(B_0 - \mu A_0)^{-1}$ by $v = (B_0 - \mu A_0)^{-1} w$ when (3.1) holds. We suppose that the resolvent is explicitly known for a particular value of μ.

We now consider two linear transformations C and D from V to W whose ranges are finite-dimensional subspaces of W. That is, C and D are of the form

$$Cv \equiv \sum_{i=1}^{k} c_i[v]w_i, \qquad Dv \equiv \sum_{i=1}^{k} d_i[v]w_i, \tag{3.2}$$

where $w_1, w_2, \cdots, w_k$ are elements of W and $c_1, \cdots, c_k, d_1, \cdots, d_k$ are linear functionals on V.

The fact that the transformation $A_0 - \mu B_0$ is one-to-one implies that μ is not an eigenvalue of the problem

$$B_0 u = \nu A_0 u.$$

We wish to know whether or not μ is an eigenvalue of the perturbed problem

$$(B_0 - D)u = \nu(A_0 + C)u \tag{3.3}$$

and what the associated eigenvectors are. The answer is given by the following theorem. (See Weinstein (1935), (1937).)

THEOREM 3.1. *The number μ in the resolvent set of* (3.1) *is an eigenvalue of the problem* (3.3) *if and only if the determinant of the matrix*

$$\delta_{ij} - d_i[(B_0 - \mu A_0)^{-1}w_j] - \mu c_i[(B_0 - \mu A_0)^{-1}w_j] \tag{3.4}$$

is zero. The multiplicity of μ as an eigenvalue of the problem (3.3) *is equal to the dimension of the null space of the matrix* (3.4), *and the corresponding eigenvectors are the elements $u = \sum_{j=1}^{k} \alpha_j(B_0 - \mu A_0)^{-1}w_j$, where α_j is any member of the null space of the matrix* (3.4).

Proof. If u is an eigenvector of the problem (3.3) corresponding to the eigenvalue $\nu = \mu$, we have

$$(B_0 - \mu A_0)u = (D + \mu C)u. \tag{3.5}$$

Hence,

$$u = (B_0 - \mu A_0)^{-1}(D + \mu C)u.$$

We see from (3.2) that u must then be a linear combination of the elements $(B_0 - \mu A_0)^{-1}w_i$:

$$u = \sum_{j=1}^{k} \alpha_j(B_0 - \mu A_0)^{-1}w_j, \tag{3.6}$$

where

$$\alpha_j = d_j[u] + \mu c_j[u].$$

Applying the functional $d_i + \mu c_i$ to both sides of (3.6), we see that

$$\alpha_i = \sum_{j=1}^{k} (d_i + \mu c_i)[(B_0 - \mu A_0)^{-1}w_j]\alpha_j. \tag{3.7}$$

Thus, the formula (3.6) gives an eigenvector if and only if α_j is in the null space of the matrix (3.4). Moreover, if $\sum_{j=1}^{k} \alpha_j(B_0 - \mu A_0)^{-1}w_j = 0$, we see from (3.7) that $\alpha_1 = \cdots = \alpha_k = 0$. Thus distinct α_j lead to distinct eigenvectors, even though the elements $w_1, \cdots, w_k$ may be linearly dependent. This completes the proof.

Remark. The determinant of the matrix (3.4) is called the *Weinstein determinant* $W(\mu)$.

If μ is an eigenvalue of finite multiplicity of the problem $B_0u = \nu A_0u$, we can still ascertain whether or not μ is an eigenvalue of the problem (3.3) and find the corresponding eigenvectors, provided we also know the eigenvectors of the original problem and its adjoint.

We shall suppose that the null space of the transformation $B_0 - \mu A_0$ is spanned by the m_0 linearly independent vectors $q_1, \cdots, q_{m_0}$. Moreover, we assume that there are n_0 linearly independent linear functionals $l_1, \cdots, l_{n_0}$ on W with the property that the problem

$$(3.8) \qquad (B_0 - \mu A_0)u = w, \qquad w \in W,$$

has a solution u if and only if

$$l_1[w] = \cdots = l_{n_0}[w] = 0$$

($l_1, \cdots, l_{n_0}$ are the eigenvectors of the adjoint problem). For each w satisfying these conditions we denote by $(B_0 - \mu A_0)^{-1}w$ some solution of (3.8). Then the general solution of the equation (3.8) is

$$u = (B_0 - \mu A_0)^{-1}w + \sum_{\nu=1}^{m_0} \beta_\nu q_\nu.$$

We again look at the eigenvalue equation for the problem (3.3) in the form (3.5). In order to have a nontrivial solution of this equation we must have

$$(3.9) \qquad l_\alpha[(D + \mu C)u] = 0, \qquad \alpha = 1, \cdots, n_0.$$

Then the solution becomes

$$(3.10) \qquad u = (B_0 - \mu A_0)^{-1}((D + \mu C)u) + \sum_{\nu=1}^{m_0} \beta_\nu q_\nu.$$

Since the w_i do not, in general, satisfy the conditions $l_1[w_i] = l_2[w_i] = \cdots = l_{n_0}[w_i] = 0$, we cannot define $(B_0 - \mu A_0)^{-1}w_i$ for all i. Hence, we cannot decompose the first term on the right of (3.10) into a linear combination of elements

$$(B_0 - \mu A_0)^{-1}w_i.$$

We circumvent this problem in the following manner.

We choose any elements $z_1, z_2, \cdots, z_{m_0}$ of W for which

$$(3.11) \qquad l_\alpha[z_\beta] = \delta_{\alpha\beta}, \qquad \alpha, \beta = 1, \cdots, m_0.$$

We now define the operator

$$(3.12) \qquad Pw \equiv w - \sum_{\alpha=1}^{n_0} l_\alpha[w]z_\alpha$$

on W. Then P is a projection onto the set of v where $l_1[v] = \cdots = l_{n_0}[v] = 0$. It follows from (3.9) that $P(D + \mu C)u = (D + \mu C)u$. Thus, we can replace (3.10) by

$$u = (B_0 - \mu A_0)^{-1}P(D + \mu C)u + \sum_{\nu=1}^{m_0} \beta_\nu q_\nu.$$

If we now define

$$\alpha_j = (d_j + \mu c_j)[u],$$

we have

$$u = \sum_{j=1}^{k} \alpha_j (B_0 - \mu A_0)^{-1} P w_j + \sum_{\nu=1}^{m_0} \beta_\nu q_\nu. \tag{3.13}$$

We apply the functional $d_i + \mu c_i$ to this equation and also write out (3.9) to find that (3.13) gives an eigenvector if and only if

$$\begin{aligned} &\alpha_i - \sum_{j=1}^{k} (d_i + \mu c_i)[(B_0 - \mu A_0)^{-1} P w_j]\alpha_j - \sum_{\nu=1}^{m_0} (d_i + \mu c_i)[q_\nu]\beta_\nu = 0, \\ &\qquad\qquad\qquad\qquad\qquad\qquad\qquad\qquad\qquad\qquad\qquad\qquad i = 1, \cdots, k, \\ &\sum_{j=1}^{k} l_\alpha[w_j]\alpha_j = 0, \qquad\qquad\qquad\qquad \alpha = 1, \cdots, n_0. \end{aligned} \tag{3.14}$$

If u as given by (3.13) is zero, we see from the first set of equations (3.14) that $\alpha_1 = \cdots = \alpha_k = 0$. Since the q_ν are linearly independent, it then follows that $\beta_1 = \cdots = \beta_{m_0} = 0$. We thus have proved the following theorem. (See Weinstein (1935), (1937).)

THEOREM 3.2. *Let μ be an eigenvalue of multiplicity m_0 of the problem $A_0 u = \mu B_0 u$, and let the range of $A_0 - \mu B_0$ be the subset of W where $l_1[w] = \cdots = l_{n_0}[w] = 0$.*

If the operator P is defined by (3.12) *in terms of elements z_α of V with the properties* (3.11), *then the null space of $B_0 - D - \mu(A_0 + C)$ with C and D defined by* (3.2) *consists of the elements* (3.13) *where $\{\alpha_j, \beta_\nu\}$ is any solution of the system* (3.14). *Distinct solutions $\{\alpha_j, \beta_\nu\}$ give distinct eigenvectors, so that the multiplicity of μ as an eigenvalue of the problem* (3.3) *is equal to the dimension of the solution space of* (3.14).

We shall now consider the special case where $V = W$ is a Hilbert space with scalar product (u, v), where A_0 is the identity operator I, and where B_0, C, and D are bounded self-adjoint operators. That is, $(u, B_0 v) = \overline{(v, B_0 u)}$, $(u, Cv) = \overline{(v, Cu)}$, and $(u, Dv) = \overline{(v, Du)}$ for all u and v in V. In addition, we shall assume that the quadratic functional (u, Cu) is positive semidefinite.

Then C and D must have the form

$$Cu = \sum_{i,j=1}^{k} (u, w_i) c_{ij} w_j, \qquad Du = \sum_{i,j=1}^{k} (u, w_i) d_{ij} w_j, \tag{3.15}$$

where the matrices c_{ij} and d_{ij} are Hermitian and c_{ij} is positive semidefinite. Thus

$$c_j[u] = \sum_{i=1}^{k} (u, w_i) c_{ij}, \qquad d_j[u] = \sum_{i=1}^{k} (u, w_i) d_{ij}. \tag{3.16}$$

The matrix (3.4) becomes

$$\delta_{ij} - \sum_{l=1}^{k} (d_{li} + \mu c_{li})((B_0 - \mu I)^{-1} w_j, w_l). \tag{3.17}$$

We shall show that under these additional assumptions one can obtain the eigenvalues of (3.3) in terms of the behavior of the Weinstein determinant

$$W(\nu) \equiv \det\left\{\delta_{ij} - \sum_{l=1}^{k} (d_{li} + \nu c_{li})((B_0 - \nu I)^{-1} w_j, w_l)\right\}. \tag{3.18}$$

It is easily seen that $W(\nu)$ is an analytic function of ν on the resolvent set of B_0 and that it has either a pole or a removable singularity at each eigenvalue of finite multiplicity. We make the following definition.

DEFINITION. The *order* $z(\mu)$ of $W(\nu)$ at μ is that integer (positive, negative, or zero) for which the product $(\nu - \mu)^{-z(\mu)}W(\nu)$ is bounded and bounded away from zero in some neighborhood of μ.

The following theorem, which is due to Aronszajn (1948), (1951), allows one to compute the eigenvalues of (3.3).

THEOREM 3.3 (Aronszajn's rule). *Let B_0 be a self-adjoint operator on a Hilbert space V with scalar product (u, v). Let C and D be defined by* (3.15) *with $w_1, \cdots, w_k$ in V, c_{ij} and d_{ij} Hermitian, and c_{ij} positive semidefinite. Denote by $m_0(\mu)$ the multiplicity (possibly zero) of μ as an eigenvalue of the problem $B_0u = \nu u$, and by $m(\mu)$ the multiplicity of μ as an eigenvalue of the problem $(B_0 - D)u = \nu(I + C)u$.*

Then if μ is not in the essential spectrum of B_0 and if $z(\mu)$ is the order at μ of the Weinstein determinant (3.18),

$$m(\mu) = m_0(\mu) + z(\mu). \tag{3.19}$$

Proof. Since the transformations are self-adjoint and (v, v) and $(v, (I + C)v)$ are positive definite, we need to consider only real values of μ.

If we choose any matrix r_{ij} with inverse s_{ij} and define the new set of elements

$$\tilde{w}_j = \sum_{i=1}^{k} r_{ij} w_i \tag{3.20}$$

and the new matrices

$$\tilde{c}_{pq} = \sum_{i,j=1}^{k} \bar{s}_{pi} c_{ij} s_{qj}, \quad \tilde{d}_{pq} = \sum_{i,j=1}^{k} \bar{s}_{pi} d_{ij} s_{qj}, \tag{3.21}$$

we can represent C and D in the form

$$Cv = \sum_{i,j=1}^{k} \tilde{c}_{ij}(v, \tilde{w}_i)\tilde{w}_j, \qquad Dv = \sum_{i,j=1}^{k} \tilde{d}_{ij}(v, \tilde{w}_i)\tilde{w}_j. \tag{3.22}$$

If we premultiply the matrix (3.17) by s_{pi} and postmultiply by r_{jl}, we obtain the matrix of the same form with $d_{ij} + \mu c_{ij}$ replaced by $\tilde{d}_{ij} + \mu\tilde{c}_{ij}$ and w_i replaced by $\tilde{w}_i$. This operation leaves the Weinstein determinant (3.18) unchanged.

In particular, if the w_i are linearly dependent, we choose r_{ij} so that the first l $\tilde{w}_i$ are linearly independent and the last $k - l$ are zero. Then the last $k - l$ columns of the matrix in (3.18) in the new coordinates are zero except for a unit matrix in the lower corner. Therefore, the determinant $W(\mu)$ is equal to the determinant which is obtained by throwing out the last $k - l$ rows and columns of the new matrix.

Thus we may assume without loss of generality that the w_i are linearly independent.

We first consider the case where μ is not an eigenvalue of the problem $B_0 u = \nu u$, so that $m_0(\mu) = 0$.

We premultiply the matrix in (3.18) by the Hermitian matrix $((B_0 - \nu I)^{-1} w_p, w_i)$ to obtain the Hermitian matrix

$$S_{pj}(\nu) \equiv ((B_0 - \nu I)^{-1} w_p, w_j) - \sum_{i,l=1}^{k} ((B_0 - \nu I)^{-1} w_p, w_i)(d_{li} + \nu c_{li})((B_0 - \nu I)^{-1} w_j, w_l). \tag{3.23}$$

It follows from the usual properties of determinants that

$$W(\nu) = \frac{\det [S_{pj}(\nu)]}{\det [(B_0 - \nu I)^{-1} w_p, w_i)]} \tag{3.24}$$

whenever the denominator is not zero.

By the corollary to Theorem 2.3, Chap. 3, we can find a matrix R with $R^*R = I$ for which $R^*S(\mu)R$ is diagonal with real entries. If R has the entries r_{ij}, the matrix R^*SR is again of the form (3.23) with w_i replaced by $\tilde{w}_i$ and $d_{ij} + \nu c_{ij}$ by $\tilde{d}_{ij} + \nu \tilde{c}_{ij}$, but it is now diagonal for $\nu = \mu$. We suppress the tildes, and we suppose that the p zero eigenvalues occur first, so that the upper left $p \times p$ submatrix is zero.

We now observe that the null space of the matrix $((B_0 - \mu I)^{-1} w_j, w_l)$ is invariant under operation by the matrix (3.17). It follows that the null space of $S_{ij}(\mu)$ is the direct sum of two disjoint spaces: the null space of the matrix (3.17) and the null space of $((B_0 - \mu I)^{-1} w_j, w_l)$. By Theorem 3.1 the null space of (3.17) has dimension $m(\mu)$. We denote by m' the dimension of the null space of $((B_0 - \mu I)^{-1} w_i, w_l)$. Then $p = m(\mu) + m'$. We arrange the matrix r_{ij} so that the first $m(\mu)$ columns of $S_{ij}(\mu)$ lie in the null space of (3.17) and the next m' columns lie in the null space of $((B_0 - \mu I)^{-1} w_j, w_l)$. Then the first $m(\mu) + m'$ rows and columns of $S(\mu)$ are zero while the remainder of $S(\mu)$ is a nonsingular diagonal matrix. Moreover, we have

$$\delta_{ij} - \sum_{l=1}^{k} (d_{il} + \mu c_{il})((B_0 - \mu I)^{-1} w_j, w_l) = 0 \quad \text{for } j = 1, \cdots, m(\mu), \quad i = 1, \cdots, k, \tag{3.25}$$

and

$$((B_0 - \mu I)^{-1} w_i, w_l) = 0 \quad \text{for } i = m(\mu) + 1, \cdots, m(\mu) + m', \quad l = 1, \cdots, k. \tag{3.26}$$

An easy computation shows that

$$\frac{d}{d\mu}((B_0 - \mu I)^{-1} w_i, w_j) = ((B_0 - \mu I)^{-1} w_i, (B_0 - \mu I)^{-1} w_j). \tag{3.27}$$

It follows from (3.25), (3.26), and (3.27) that

$$\frac{d}{d\mu}S_{ij}(\mu) = \begin{cases} -((B_0 - \mu I)^{-1}w_i, (B_0 - \mu I)^{-1}w_j) \\ \quad - \sum_{p,q=1}^{k} ((B_0 - \mu I)^{-1}w_i, w_p)c_{pq}((B_0 - \mu I)^{-1}w_q, w_j) \\ \qquad\qquad \text{for } i,j \leqq m(\mu), \\ 0 \text{ for } i \leqq m(\mu), m(\mu) < j \leqq m(\mu) + m', \\ 0 \text{ for } j \leqq m(\mu), m(\mu) < i \leqq m(\mu) + m', \\ ((B_0 - \mu I)^{-1}w_i, (B_0 - \mu I)^{-1}w_j) \text{ for } m(\mu) \leqq i,j \leqq m(\mu) + m'. \end{cases}$$

We rearrange the matrix r_{ij} so that the matrix $(d/d\mu)(S_{ij}(\mu))$ with $i,j \leqq m(\mu) + m'$ is diagonal and recall that the matrix c_{pq} is positive semidefinite. Then the first $m(\mu)$ diagonal entries are negative while the next m' are positive. Since none of them is zero, we conclude that for ν near μ the determinant of $S_{ij}(\nu)$ is equal to $(\nu - \mu)^{m(\mu)+m'}$ times a nonzero constant. That is, $\det[S_{ij}(\nu)]$ has a zero of order $m(\mu) + m'$ at μ.

A similar argument shows that $\det((B_0 - \mu I)^{-1}w_i, w_j)$ has a zero or order m' at μ. Hence the order $z(\mu)$ of $W(\nu)$ at μ is equal to $m(\mu)$. Thus the theorem is verified when $m_0(\mu) = 0$.

Suppose now that the null space of $B_0 - \mu I$ is of dimension $m_0(\mu) > 0$ and that it is spanned by the elements $q_1, \cdots, q_{m_0}$. (We have written m_0 for $m_0(\mu)$.) We normalize these q_α so that

$$(q_\alpha, q_\beta) = \delta_{\alpha\beta}.$$

We define the orthogonal projection

$$Qf = \sum_{\alpha=1}^{m_0} \overline{(q_\alpha, f)}q_\alpha.$$

We note that the equation $(B_0 - \nu I)u = (D + \nu C)u$ is equivalent to

$$(B_0 + Q - \nu I)u = (D + Q + \nu C)u. \tag{3.28}$$

Thus, we may replace B_0 by $B_0 + Q$ and D by the operator $D + Q$, which again has finite-dimensional range. It is easily verified that for any f in W and ν sufficiently near μ,

$$(B_0 + Q - \nu I)^{-1}f = (B_0 - \nu I)^{-1}(f - Qf) + \frac{1}{1 + \mu - \nu}Qf.$$

In particular, we find that μ is in the resolvent set, so that the multiplicity $m(\mu)$ of μ as an eigenvalue of (3.28) is equal to the order of the analogue of $W(\nu)$ for this problem. Since

$$Qu = \sum_{\alpha=1}^{m_0} \overline{(q_\alpha, u)}q_\alpha,$$

we must append the elements q_α to the w_i in order to obtain the range of the perturbation. We also add the linear functionals $\overline{(q_\alpha, u)}$ to the set of d_i. In this way we see that $m(\mu)$ is equal to the order of

$$\det\left(\begin{array}{c|c} \delta_{ij} - (d_i + \nu c_i)\left[(B_0 - \nu I)^{-1}(w_j - Qw_j) + \dfrac{1}{1+\mu-\nu}Qw_j\right] & -\dfrac{1}{1+\mu-\nu}(d_i + \nu c_i)[q_\beta] \\ \hline -\dfrac{1}{1+\mu-\nu}\overline{(q_\alpha, w_j)} & \dfrac{\mu-\nu}{\mu-\nu+1}\delta_{\alpha\beta} \end{array}\right)$$

$$i, j = 1, \cdots, k, \quad \alpha, \beta = 1, \cdots, m_0,$$

at $\nu = \mu$. If we add the sum of $(\mu - \nu)^{-1}\overline{(q_\beta, w_j)}$ times the $(k + \beta)$th column to the jth column, this determinant becomes

$$\det\left(\begin{array}{c|c} \delta_{ij} - (d_i + \nu c_i)[(B_0 - \nu I)^{-1}w_j] & -\dfrac{1}{\mu-\nu}(d_i + \nu c_i)[q_\beta] \\ \hline 0 & \dfrac{\mu-\nu}{\mu-\nu+1}\delta_{\alpha\beta} \end{array}\right)$$

$$= \left(\frac{\mu-\nu}{\mu-\nu-1}\right)^{m_0(\mu)} W(\nu).$$

Therefore its order $m(\mu)$ is equal to $z(\mu) + m_0(\mu)$, which proves the theorem.

Remark 1. In the case $C = 0$ Theorem 3.3 has been extended to certain infinite-dimensional not necessarily Hermitian perturbations by Kuroda (1961). (See also Kato (1966, IV.6).)

Remark 2. While Theorem 3.3 is of great theoretical interest and is aesthetically more pleasing than Theorems 3.1 and 3.2, it is not necessarily easier to use for computational purposes. Moreover, it does not give the eigenvectors, as Theorems 3.1 and 3.2 do.

Remark 3. It is easily seen that if μ is in the essential spectrum of B_0, it is also in the essential spectrum of $(I + C)^{-1}(B_0 - D)$.

4. The method of A. Weinstein. Suppose that for the problem

$$Bu = \mu A u, \qquad u \in V^{(0)}, \tag{4.1}$$

the eigenvalues $\mu_1^{(0)} \geqq \mu_2^{(0)} \geqq \cdots$ and their corresponding eigenvectors $u_1^{(0)}$, $u_2^{(0)}, \cdots$ are known. We wish to approximate the eigenvalues $\mu_1 \geqq \mu_2 \geqq \cdots$ of the problem

$$Bu = \mu A u, \qquad u \in V,$$

where V is a linear subspace of the completion $V_{\mathscr{A}}^{(0)}$ of $V^{(0)}$. By the first monotonicity principle we know that

$$\mu_n \leqq \mu_n^{(0)}$$

so that we have upper bounds for the μ_n. Improvable lower bounds can be obtained from the Rayleigh–Ritz method. We wish to improve the upper bounds $\mu_n^{(0)}$.

If the completion $V_{\mathscr{A}}$ coincides with $V_{\mathscr{A}}^{(0)}$, we have $\mu_n = \mu_n^{(0)}$ and no improvement is needed. If this is not the case, there is a nonempty linear vector space W which is the orthogonal complement of V in $V_{\mathscr{A}}^{(0)}$:

$$W = \{w \in V_{\mathscr{A}}^{(0)} | \mathscr{A}(v, w) = 0 \text{ for all } v \in V\}.$$

We choose some linearly independent elements $\phi_1, \phi_2, \cdots, \phi_k$ in W, and we let

$$V^{(k)} = \{v \in V_{\mathscr{A}}^{(0)} | \mathscr{A}(v, \phi_i) = 0, i = 1, \cdots, k\}.$$

Clearly $V_{\mathscr{A}}^{(0)} \supset V^{(k)} \supset V$. Hence by the first monotonicity principle, the eigenvalues $\tilde{\mu}_n$ which are defined as the successive maxima of $\mathscr{B}(v, v)/\mathscr{A}(v, v)$ on $V^{(k)}$ satisfy

$$\mu_n^{(0)} \geqq \tilde{\mu}_n \geqq \mu_n.$$

Thus $\tilde{\mu}_n$ is in general a better upper bound than $\mu_n^{(0)}$. We must, however, find the eigenvalues $\tilde{\mu}_n$ before we can use them as upper bounds.

Let $P^{(k)}$ be the orthogonal projection onto the space $V^{(k)}$. For simplicity, we assume that the ϕ_i are orthonormal in the sense that $\mathscr{A}(\phi_i, \phi_j) = \delta_{ij}$. Then

$$P^{(k)}u = u - \sum_i^k \mathscr{A}(u, \phi_i)\phi_i. \tag{4.2}$$

We define the sesquilinear functional $\mathscr{B}^{(k)}$ and the operator $T^{(k)}$ by

$$\mathscr{B}^{(k)}(u, v) \equiv \mathscr{B}(P^{(k)}u, P^{(k)}v) \equiv \mathscr{A}(u, T^{(k)}v). \tag{4.3}$$

The operator $T^{(k)}$ clearly has the eigenvalue $\mu = 0$ with the corresponding eigenvectors $\phi_1, \phi_2, \cdots, \phi_k$. On the orthogonal complement of these eigenvectors we have $P^{(k)}v = v$ so that $\mathscr{B}^{(k)} = \mathscr{B}$. Therefore the eigenvalues of $T^{(k)}$ are exactly the $\tilde{\mu}_n$ together with a k-fold eigenvalue at $\mu = 0$.

If we substitute (4.2) in the definition (4.3) of $T^{(k)}$ we find that

$$T^{(k)}v = Tv - \sum_{i=1}^k \mathscr{A}(v, \phi_i)T\phi_i - \sum_{i=1}^k \mathscr{B}(v, \phi_i)\phi_i + \sum_{i,j=1}^k \mathscr{A}(v, \phi_i)\mathscr{B}(\phi_i, \phi_j)\phi_j.$$

Thus $T^{(k)}$ is of the form $B_0 - D$ where $B_0 = T$ and the range of D consists of the linear combinations of the ϕ_i and $T\phi_i$.

We thus have a problem of the form (3.3) with $C = 0$, $A_0 = I$, and $(u, v) \equiv \mathscr{A}(u, v)$.

The set of w_i consists of $\phi_1, \phi_2, \cdots, \phi_k, T\phi_1, \cdots, T\phi_k$. A short computation which uses the identity $(T - \mu I)^{-1}T = I + \mu(T - \mu I)^{-1}$ shows that the matrix (3.17) takes the form

$$\left(\begin{array}{c|c} \sum_{l=1}^k \mathscr{A}(\phi_l, T\phi_i - \mu\phi_i)\mathscr{A}((T-\mu I)^{-1}\phi_j, \phi_l) & -\mu\delta_{in} + \mu \sum_{l=1}^k \mathscr{A}(\phi_l, T\phi_i - \mu\phi_i)\mathscr{A}((T-\mu I)^{-1}\phi_n, \phi_l) \\ \hline -\mathscr{A}((T-\mu I)^{-1}\phi_j, \phi_m) & -\mu\mathscr{A}((T-\mu I)^{-1}\phi_n, \phi_m) \end{array}\right).$$

If we premultiply the lower half of this matrix by $\mathscr{A}(\phi_m, T\phi_i - \mu\phi_i)$ and add the result to the upper half, we obtain the matrix

$$\left(\begin{array}{c|c} 0 & -\mu\delta_{in} \\ \hline -\mathscr{A}((T-\mu I)^{-1}\phi_j, \phi_m) & -\mu\mathscr{A}((T-\mu I)^{-1}\phi_n, \phi_m) \end{array}\right). \tag{4.4}$$

If $\mu \neq 0$, the last k coefficients vanish on the null space of this matrix.

Theorem 3.1 now shows that a nonzero μ which is not an eigenvalue of T is an eigenvalue of $T^{(k)}$ if and only if the matrix $\mathscr{A}((T - \mu I)^{-1}\phi_i, \phi_j)$ is singular, and that the eigenvectors are of the form

$$u = \sum_{j=1}^{k} \alpha_j (T - \mu I)^{-1}\phi_j, \tag{4.5}$$

where α_j lies in the null space of this matrix. The condition that α_j lie in this null space just means that u is in the space $V^{(k)}$.

A. Weinstein (1935), (1937) obtained this result together with Theorem 3.2 to determine the eigenvalues of $T^{(k)}$, which then give upper bounds for the corresponding eigenvalues μ_n of the restriction of T to $V_{\mathscr{A}}$. In order to determine which upper bound is for which eigenvalue μ_n it is essential to obtain all the eigenvalues of $T^{(k)}$ including those which coincide with eigenvalues of T on $V^{(0)}$, and to determine their proper multiplicities.

In order to apply Theorem 3.3 we note that the determinant of the matrix (4.4) is

$$W(\mu) = (-1)^k \mu^k \det \mathscr{A}((T - \mu I)^{-1}\phi_i, \phi_j). \tag{4.6}$$

Its order $z(\mu)$ agrees with that of $\det \mathscr{A}((T - \mu I)^{-1}\phi_i, \phi_j)$ except at $\mu = 0$. At $\mu = 0$ the factor μ^k gives the multiplicity k, which corresponds to the null vectors $\phi_1, \phi_2, \cdots, \phi_k$ of $T^{(k)}$. These elements are not eigenvectors of the restriction of T to $V^{(k)}$. Finally, we note that the order of the function (4.6) is left unchanged if the ϕ_i are replaced by k linearly independent linear combinations. Therefore they need not be orthonormal. Thus Theorem 3.3 becomes the following theorem.

THEOREM 4.1 (Weinstein method). *Let $\phi_1, \phi_2, \cdots, \phi_k$ be any linearly independent elements of $V_{\mathscr{A}}^{(0)}$ which are orthogonal to V. Let $m_0(\mu)$ and $m(\mu)$ be the multiplicities of μ as an eigenvalue of T and the restriction of T to $V^{(k)}$, respectively. Let $z(\mu)$ be the order at μ of the function*

$$W(\nu) \equiv \det \mathscr{A}((T - \nu I)^{-1}\phi_i, \phi_j). \tag{4.7}$$

Then

$$m(\mu) = m_0(\mu) + z(\mu).$$

The eigenvalues $\tilde{\mu}_1 \geqq \tilde{\mu}_2 \geqq \cdots$ so obtained are upper bounds for the corresponding eigenvalues of the restriction of $\mathscr{B}(u, u)/\mathscr{A}(u, u)$ to $V_{\mathscr{A}}$.

Remark. The determinant (4.7) is the determinant that was introduced by A. Weinstein.

There are several means of expressing the entries of the determinant $W(\mu)$. If we have an explicit solution operator $u = (B - \mu A)^{-1}f$ for the problem

$$Bu - \mu Au = f, \qquad u \in V^{(0)},$$

then

$$(T - \mu I)^{-1}\phi_i = (B - \mu A)^{-1}A\phi_i.$$

Thus,

$$\mathscr{A}(\phi_i, (T - \mu I)^{-1}\phi_j) = ((B - \mu A)^{-1}A\phi_i, A\phi_j). \tag{4.8}$$

If the quadratic functional $\mathscr{B}(u, u)$ is completely continuous with respect to $\mathscr{A}(u, u)$ and if the eigenvalues $\mu_1 \geqq \mu_2 \geqq \cdots$ and $-\mu_1^{(-1)} \leqq -\mu_2^{(-)} \leqq \cdots$ and the corresponding orthonormal eigenvectors $u_n(0)$ and $u_n^{(0,-)}$ are known, we can use Parseval's equations (3.4), Chap. 3, to show that

$$\begin{aligned}\mathscr{A}(\phi_i, (T - \mu I)^{-1}\phi_j) &= \sum \frac{\mu_n \mathscr{A}(\phi_j, u_n^{(0)})\overline{\mathscr{A}(\phi_j, u_n^{(0)})}}{\mu(\mu_n - \mu)} \\ &+ \sum \frac{\mu_n^{(-)}\mathscr{A}(\phi_i, u_n^{(0,-)})\overline{\mathscr{A}(\phi_j, u_n^{(0,-)})}}{\mu(\mu_n^{(-)} + \mu)} - \frac{1}{\mu}\mathscr{A}(\phi_i, \phi_j).\end{aligned} \tag{4.9}$$

We observe that the separation theorem (Theorem 9.1, Chap. 3) shows that $\tilde{\mu}_n$ must satisfy the inequality

$$\tilde{\mu}_n \geqq \mu_{n+k}^{(0)}$$

as well as $\tilde{\mu}_n \geqq \mu_n$. It can be shown (Weinberger (1952b)) that when $\mathscr{B}$ is completely continuous, these are the only limitations on how small the $\tilde{\mu}_n$ can be made. That is, for each n there are elements $\phi_1, \cdots, \phi_k$ for which one of these inequalities becomes an equality.

If $\{\phi_i\}$ is any infinite sequence of elements whose linear combinations are dense in W and if T is completely continuous, then it was shown by Aronszajn and Weinstein (1942) and Aronszajn (1951) that the upper bound $\tilde{\mu}_n$ obtained by applying the Weinstein method with $\phi_1, \cdots, \phi_k$ converges to μ_n as $k \to \infty$.

Detailed discussions of the Weinstein method are to be found in the books of Gould (1966) and Weinstein and Stenger (1972).

5. The method of N. Aronszajn. We now consider a problem where the second monotonicity principle applies, and seek to improve the corresponding bounds.

We suppose that the quadratic functionals $\mathscr{A}(u, u)$ and $\mathscr{B}(u, u)$ can be written in the form

$$\begin{aligned}\mathscr{A}(u, u) &= \mathscr{A}_0(u, u) + \mathscr{A}'(u, u), \\ \mathscr{B}(u, u) &= \mathscr{B}_0(u, u) - \mathscr{B}'(u, u).\end{aligned} \tag{5.1}$$

The functionals $\mathscr{A}_0$ and $\mathscr{B}_0$ are assumed to have the same properties as $\mathscr{A}$ and $\mathscr{B}$. In addition, we suppose that we know the eigenvalues $\mu_1^{(0)} \geqq \mu_2^{(0)} \geqq \cdots$, the eigenvectors $u_1^{(0)}, u_2^{(0)}, \cdots$, and the resolvent $(T_0 - \mu I)^{-1}$ of the operator T_0 defined by

$$\mathscr{B}_0(v, u) = \mathscr{A}_0(v, T_0 u). \tag{5.2}$$

We further suppose that the functionals $\mathscr{A}'(u, u)$ and $\mathscr{B}'(u, u)$ are positive semi-definite:

$$\mathscr{A}'(u,u) \geqq 0, \qquad \mathscr{B}'(u,u) \geqq 0. \tag{5.3}$$

Finally we assume that there are known not necessarily bounded linear transformations M and N from V to $V_{\mathscr{A}_0}$ with the properties that

$$\mathscr{A}'(u,v) = \mathscr{A}_0(u, Mv), \qquad \mathscr{B}'(u,v) = \mathscr{A}_0(u, Nv) \tag{5.4}$$

for all u and v in V.

We wish to approximate the eigenvalues $\mu_1 \geqq \mu_2 \geqq \cdots$ of the Rayleigh quotient $\mathscr{B}(u, u)/\mathscr{A}(u, u)$. By the second monotonicity principle we have

$$\mu_n \leqq \mu_n^{(0)}$$

when $\mu_n^{(0)} \geqq 0$. We wish to improve these bounds.

Aronszajn's method is based on the following idea. Let $\zeta_1, \zeta_2, \cdots, \zeta_a$ be a set of elements of V with the property that the matrix $\mathscr{A}'(\zeta_i, \zeta_j)$ is nonsingular (unless $a = 0$). We define the projection operator

$$Q_1 u = \sum_{i=1}^{a} c_i \zeta_i, \tag{5.5}$$

where the c_i are to be chosen so that

$$\mathscr{A}'(Q_1 u - u, \zeta_j) = 0, \qquad j = 1, \cdots, a. \tag{5.6}$$

Similarly, we choose $\psi_1, \cdots, \psi_b$ so that $\mathscr{B}'(\psi_i, \psi_j)$ is nonsingular (unless $b = 0$), and define

$$Q_2 u = \sum_{\alpha=1}^{b} d_\alpha \psi_\alpha \tag{5.7}$$

so that

$$\mathscr{B}'(Q_2 u - u, \psi_\beta) = 0, \qquad \beta = 1, \cdots, b. \tag{5.8}$$

It is easily seen that

$$\begin{aligned} 0 &\leqq \mathscr{A}'(Q_1 u, Q_1 u) \leqq \mathscr{A}'(u, u), \\ 0 &\leqq \mathscr{B}'(Q_2 u, Q_2 u) \leqq \mathscr{B}'(u, u). \end{aligned} \tag{5.9}$$

By the second monotonicity principle the eigenvalues $\tilde{\mu}_1 \geqq \tilde{\mu}_2 \geqq \cdots$ of the Rayleigh quotient

$$\frac{\mathscr{B}_0(u,u) - \mathscr{B}'(Q_2 u, Q_2 u)}{\mathscr{A}_0(u,u) + \mathscr{A}'(Q_1 u, Q_1 u)} \tag{5.10}$$

satisfy the inequalities

$$\mu_n \leqq \tilde{\mu}_n \leqq \mu_n^{(0)}$$

whenever $\tilde{\mu}_n \geqq 0$. Thus, the $\tilde{\mu}_n$ give improved upper bounds, provided they can be found.

Because of (5.2) and (5.4) the Rayleigh quotient (5.10) can be written in the form

$$\frac{\mathscr{A}_0(u, T_0u - NQ_2u)}{\mathscr{A}_0(u, u + MQ_1u)}.$$

Thus its eigenvalues are those of the eigenvalue problem

$$T_0u - NQ_2u = \tilde{\mu}(u + MQ_1u). \tag{5.11}$$

The upper bounds $\tilde{\mu}_n$ can now be determined by applying Theorems 3.1 and 3.2 or Theorem 3.3. The Weinstein determinant (3.18) for this problem is

$$W(\nu) = \begin{vmatrix} \mathscr{A}_0((T_0 - \nu I)^{-1}M\zeta_i - \nu^{-1}\zeta_i, M\zeta_j) & \mathscr{A}_0((T_0 - \nu I)^{-1}M\zeta_i, N\psi_\beta) \\ \mathscr{A}_0((T_0 - \nu I)^{-1}N\psi_\alpha, M\zeta_j) & \mathscr{A}_0((T_0 - \nu I)^{-1}N\psi_\alpha - \psi_\alpha, N\psi_\beta) \end{vmatrix}. \tag{5.12}$$

Theorem 3.3 yields the next theorem.

THEOREM 5.1 (Aronszajn method). *Suppose that the nonnegative eigenvalues $\mu_1^{(0)} \geqq \mu_2^{(0)} \geqq \cdots$ of the Rayleigh quotient $\mathscr{B}_0(u, u)/\mathscr{A}_0(u, u)$ and the resolvent $(T_0 - \mu I)^{-1}$ for $\mu \geqq 0$ are known. Choose elements $\zeta_1, \cdots, \zeta_a$ of V so that the matrix $\mathscr{A}_0(\zeta_i, M\zeta_j)$ is nonsingular and elements $\psi_1, \cdots, \psi_b$ so that the matrix $\mathscr{A}_0(\psi_\alpha, N\psi_\beta)$ is nonsingular. The nonnegative eigenvalues $\tilde{\mu}_1 \geqq \tilde{\mu}_2 \geqq \cdots$ of the Rayleigh quotient* (5.10) *can be found by applying Theorem* 3.3 *with the Weinstein determinant* (5.12). *These nonnegative $\tilde{\mu}_i$ are upper bounds for the corresponding eigenvalues $\mu_1 \geqq \mu_2 \geqq \cdots$ of the Rayleigh quotient $[\mathscr{B}_0(u, u) - \mathscr{A}_0(u, Nu)]/[\mathscr{A}_0(u, u) + \mathscr{A}_0(u, Mu)]$, provided the quadratic functionals $\mathscr{A}_0(u, Mu)$ and $\mathscr{A}_0(u, Nu)$ are positive semidefinite.*

Remark 1. We see from the remark after Theorem 8.1, Chap. 3, that if $M = 0$, all the $\tilde{\mu}_i$, rather than just the nonnegative ones, furnish upper bounds for the corresponding μ.

Remark 2. It follows from the second monotonicity principle that the $\tilde{\mu}_i$ do not increase when more ζ_i or ψ_α are used.

Remark 3. It was shown by Aronszajn (1951) that if $\{\zeta_i\}$ and $\{\psi_\alpha\}$ are infinite sequences whose linear combinations are dense in $V_{\mathscr{A}_0}$ and if $\mathscr{B}_0(u, u)$ is completely continuous with respect to $\mathscr{A}_0(u, u) + \mathscr{A}_0(u, Mu)$, then for each i the upper bound $\tilde{\mu}_i$ obtained by using $\zeta_1, \cdots, \zeta_a, \psi_1, \cdots, \psi_b$ converges to μ_i as $a \to \infty$ and $b \to \infty$.

In order to unify the problem of finding error bounds we shall now show that the Aronszajn method can be derived from the Weinstein method.

We define a new space whose elements are triples $\{u, v, w\}$ with u, v, and w in V. On this space we introduce the quadratic functional

$$\mathscr{C}(\{u, v, w\}, \{u, v, w\}) = \kappa\mathscr{A}_0(u, u) - \mathscr{B}_0(u, u) + \kappa\mathscr{A}'(v, v) + \mathscr{B}'(w, w), \tag{5.13}$$

where κ is any positive number greater than $\mu_1^{(0)}$. By hypothesis, $\mathscr{C}$ is positive semidefinite. We assume for the moment that $\mathscr{A}'$ and $\mathscr{B}'$ are positive definite, so that $\mathscr{C}$ is positive definite. We then complete this space of triplets to a Hilbert space H_0 with norm $\mathscr{C}(\{u, v, w\}, \{u, v, w\})^{1/2}$.

We now let

$$\mathscr{D}(\{u, v, w\}, \{u, v, w\}) = \mathscr{B}_0(u, u) - \mathscr{B}'(w, w). \tag{5.14}$$

We note that for any $\{u, v, w\}$ and $\{u', v', w'\}$ in H_0

$$\mathscr{C}(\{u, v, w\}, \{u', v', w'\}) = \mathscr{A}_0(u, \kappa u' - T_0 u') + \mathscr{A}_0(v, \kappa M v') + \mathscr{A}_0(w, N w'),$$

$$\mathscr{D}(\{u, v, w\}, \{u', v', w'\}) = \mathscr{A}_0(u, T_0 u') - \mathscr{A}_0(w, N w').$$

Therefore,

$$\mathscr{D}(\{u, v, w\}, \{u', v', w'\}) = \mathscr{C}(\{u, v, w\}, \mathscr{S}\{u', v', w'\}),$$

where the operator $\mathscr{S}$ is defined by

$$\mathscr{S}\{u', v', w'\} = \{(\kappa I - T_0)^{-1} T_0 u', 0, -w'\}.$$

The successive maxima $\nu_1^{(0)} \geqq \nu_2^{(0)} \geqq \cdots$ of the Rayleigh quotient $\mathscr{D}(\{u, v, w\}, \{u, v, w\})/\mathscr{C}(\{u, v, w\}, \{u, v, w\})$ thus give the highest eigenvalues of the operator $\mathscr{S}$.

It is easily seen that this operator has the eigenvalue 0 with all triplets of the form $\{0, v, 0\}$ as eigenvectors, the eigenvalue -1 with the triplets $\{0, 0, w\}$ as eigenvectors, and the eigenvalues $\mu_i^{(0)}/(\kappa - \mu_i^{(0)})$ with the eigenvectors $\{u_i^{(0)}, 0, 0\}$.

It is easily verified that for ν not an eigenvalue

$$(\mathscr{S} - \nu I)^{-1}\{u, v, w\} = \left\{\frac{1}{1+\nu}\left(T_0 - \frac{\kappa\nu}{1+\nu} I\right)^{-1} (\kappa I - T_0) u, -\frac{1}{\nu} v, -\frac{1}{1+\nu} w\right\}. \tag{5.15}$$

On the subspace H of H_0 which consists of $\{u, v, w\}$ with $v = w = u$ we have

$$\frac{\mathscr{D}(\{u, u, u\}, \{u, u, u\})}{\mathscr{C}(\{u, u, u\}, \{u, u, u\})} = \frac{\mathscr{B}_0(u, u) - \mathscr{B}'(u, u)}{\kappa \mathscr{A}_0(u, u) - \mathscr{B}_0(u, u) + \kappa \mathscr{A}'(u, u) + \mathscr{B}'(u, u)}$$

$$= \frac{\mathscr{B}(u, u)}{\kappa \mathscr{A}(u, u) - \mathscr{B}(u, u)}.$$

The successive maxima $\nu_1 \geqq \nu_2 \geqq \cdots$ of this quotient are just $\nu_i = \mu_i/(\kappa - \mu_i)$, where the μ_i are the eigenvalues which we wish to bound. We thus have reduced our problem to one to which the Weinstein method is applicable: To find the successive maxima of $\mathscr{D}(\{u, v, w\}, \{u, v, w\})/\mathscr{C}(\{u, v, w\}, \{u, v, w\})$ on the subspace H in terms of the known eigenvalues, eigenvectors, and inverse operator (5.15) on H_0. The orthogonal complement in H_0 of H consists of all triplets of the form

$$\{-(\kappa I - T_0)^{-1}(M\zeta + N\psi), \kappa^{-1}\zeta, \psi\}.$$

To apply the Weinstein method, we merely choose two collections of linearly independent elements $\{\zeta_1, \cdots, \zeta_a\}$ and $\{\psi_1, \cdots, \psi_b\}$ of V. We compute the elements of the matrix in (4.7) when the ϕ_i are replaced by the elements $\{-(\kappa I - T_0)^{-1} M \zeta_i, \kappa^{-1}\zeta_i, 0\}$ and $\{-(\kappa I - T_0)^{-1} N \psi_\alpha, \psi_\alpha, 0\}$, and $(T_0 - \nu I)^{-1}$ is replaced by $(\mathscr{S} - \nu I)^{-1}$ as given in (5.15). The Weinstein determinant becomes the block determinant

$$\begin{vmatrix} \mathscr{A}_0\left(\frac{1}{1+\nu}\left(T_0 - \frac{\kappa\nu}{1+\nu}I\right)^{-1} M\zeta_i - \frac{1}{\kappa\nu}\zeta_i, M\zeta_j\right) & \frac{1}{1+\nu}\mathscr{A}_0\left(\left(T_0 - \frac{\kappa\nu}{1+\nu}I\right)^{-1} M\zeta_i, N\psi_\beta\right) \\ \frac{1}{1+\nu}\mathscr{A}_0\left(\left(T_0 - \frac{\kappa\nu}{1+\nu}I\right)^{-1} N\psi_\alpha, M\zeta_j\right) & \frac{1}{1+\nu}\mathscr{A}_0\left(N\psi_\alpha, \left(T_0 - \frac{\kappa\nu}{1+\nu}I\right)^{-1} N\psi_\beta - \psi_\beta\right) \end{vmatrix}$$

We now recall that the desired eigenvalues μ_i of the projection of T are related to the eigenvalues ν_i of the projection of $\mathscr{S}$ by $\nu_i = \mu_i/(\kappa - \mu_i)$. We introduce the new variable μ by the equation $\nu = \mu/(\kappa - \mu)$. Multiplying the above matrix by $1 + \nu$, we obtain the Weinstein determinant (5.12), and hence Theorem 5.1.

We have derived this result under the hypothesis that $\mathscr{A}'(u, u)$ and $\mathscr{B}'(u, u)$ are positive definite. If they are only semidefinite, we simply define two triplets $\{u, v, w\}$ and $\{u', v', w'\}$ to be equivalent if $u = u'$ and $\mathscr{A}'(v - v', v - v') = \mathscr{B}'(v - v', v - v') = 0$. Then $\mathscr{C}(\{u, v, w\}, \{u, v, w\})^{1/2}$ is a norm on the space of equivalence classes and we may complete this space to a Hilbert space H_0. Since both $\mathscr{C}(\{u, v, w\}, \{u, v, w\})$ and $\mathscr{D}(\{u, v, w\}, \{u, v, w\})$ are constant on each equivalence class, all our results are still valid. One needs only to require that the elements $M\zeta_\alpha$ and the elements $N\psi_i$ be linearly independent. Otherwise the Weinstein determinant is identically zero. This requirement is equivalent to asking that the matrices $\mathscr{A}'(\zeta_i, \zeta_j)$ and $\mathscr{B}'(\psi_\alpha, \psi_\beta)$ be nonsingular.

It has been shown by Fichera (1965) that when $\mathscr{B}(u, u)$ is positive definite, the Weinstein method can also be derived from the Aronszajn method.

6. Bazley's choice. The Aronszajn method presents grave practical difficulties in many problems. In the first place, if $\mathscr{A}'(u, u)$ and $\mathscr{B}'(u, u)$ are given, it is not always easy to find the operators M and N defined by (5.4). In the second place, there are very few problems in which the resolvent operator $(T_0 - \mu I)^{-1}$ is known except possibly as an infinite series. In the next two sections we shall show how these difficulties may be overcome in some special situations.

We first specialize by assuming that $\mathscr{B}(u, v) = \mathscr{B}_0(u, v)$ so that $\mathscr{B}'(u, v) = 0$. We do not need to use any ψ_α in this case.

The idea of Bazley (1959), (1961) is to choose the elements $\zeta_1, \zeta_2, \cdots$ in such a way that

$$M\zeta_i = u_i^{(0)}, \tag{6.1}$$

the ith eigenfunction of T_0. Recalling the definition (5.4) of M, and the fact that $u_i^{(0)}$ is an eigenvalue, we see that (6.1) can be written as

$$\mathscr{B}(v, u_i^{(0)}) = \mu_i^{(0)}\mathscr{A}'(v, \zeta_i) \quad \text{for all } v \in V. \tag{6.2}$$

Whether or not this equation can be solved depends, of course, on the Hermitian functionals $\mathscr{B}$ and $\mathscr{A}'$. If

$$\mathscr{B}(v, w) = \int_D b(x)v(x)\overline{w(x)}\,dx$$

and

$$\mathscr{A}'(v, w) = \int a(x)v(x)\overline{w(x)}\,dx,$$

where a and b are real and $a(x)$ is bounded away from zero, then (6.2) is solved by

$$\zeta_i = \frac{b}{\mu_i^{(0)}a}u_i^{(0)}.$$

Once this solution has been found, we observe that by (6.1) the entries in (5.12) become

$$\frac{1}{\mu_i^{(0)} - \mu}\delta_{ij} - \frac{1}{\mu}\mathscr{A}'(\zeta_i, \zeta_j).$$

In fact, we can write the equation (5.11) as

$$T_0 u = \tilde{\mu}\left(u + \sum_{i,j=1}^{a} \mathscr{A}_0(u, u_i^{(0)})R_{ij}u_j^{(0)}\right),$$

where R_{ij} is the inverse matrix of $\mathscr{A}'(\zeta_i, \zeta_j)$.

Any eigenvector $u_l^{(0)}$ of T_0 with $l > a$ is also an eigenvector of this problem with the same eigenvalue $\mu_l^{(0)}$. Thus only the first a eigenvalues are shifted. However, one must be careful because some of them may be shifted below $\mu_{a+1}^{(0)}$.

The additional eigenvalues are to be found from the matrix eigenvalue problem

$$\det\left[\mu_i^{(0)}\mathscr{A}'(\zeta_i, \zeta_j) - \mu\{\delta_{ij} + \mathscr{A}'(\zeta_i, \zeta_j)\}\right] = 0.$$

This idea has been applied with great success to problems in quantum mechanics by Bazley and Fox (1961), (1963).

A similar idea can be applied to the case where $\mathscr{B}'(u, u) \not\equiv 0$ if we can solve the problem

$$N\psi_\alpha = u_\alpha^{(0)}, \qquad \alpha = 1, \cdots, b,$$

or

$$\mu_\alpha^{(0)}\mathscr{B}'(v, \psi_\alpha) = \mathscr{B}(v, u_\alpha^{(0)}) \quad \text{for all } v \in V.$$

7. Truncation. In most problems the resolvent $(T - \mu I)^{-1}$ in the Weinstein method or $(T_0 - \mu I)^{-1}$ in the Aronszajn method can only be expressed as an infinite series. This fact produces serious difficulties. In an actual computation the infinite series must be replaced by a finite one. This clearly affects the location of the zeros of the determinant, and consequently one does not know whether the computed eigenvalues are actually upper bounds for the μ_n.

We observe that if T has the eigenvalues $\mu_1^{(0)} \geqq \mu_2^{(0)} \geqq \cdots \geqq \mu_P^{(0)}$ with the corresponding orthonormal eigenvectors $u_1^{(0)}, u_2^{(0)}, u_3^{(0)}, \cdots, u_P^{(0)}$, then by the recursive definition of the eigenvalues,

$$\mathscr{B}\left(u - \sum_{i=1}^{P} \mathscr{A}(u, u_i^{(0)})u_i^{(0)}, u - \sum_{i=1}^{P} \mathscr{A}(u, u_i^{(0)})u_i^{(0)}\right)$$
$$\leqq \mu_{P+1}\mathscr{A}\left(u - \sum_{i=1}^{P} \mathscr{A}(u, u_i^{(0)})u_i^{(0)}, u - \sum_{i=1}^{P} \mathscr{A}(u, u_i^{(0)})u_i^{(0)}\right),$$

or, expanding the quadratic functionals,

$$\mathscr{B}(u, u) - \sum_{i=1}^{P} \mu_i|\mathscr{A}(u, u_i^{(0)})|^2 \leqq \mu_{P+1}\left[\mathscr{A}(u, u) - \sum_{i=1}^{P} |\mathscr{A}(u, u_i^{(0)})|^2\right].$$

Transposing, we find that

$$\mathscr{B}(u, u) \leqq \sum_{i=1}^{P} (\mu_i - \mu_{P+1})|\mathscr{A}(u, u_i^{(0)})|^2 + \mu_{P+1}\mathscr{A}(u, u). \tag{7.1}$$

Thus, if we define

$$\tilde{\mathscr{B}}(u, v) = \sum_{i=1}^{P} (\mu_i - \mu_{P+1})\mathscr{A}(u, u_i^{(0)})\overline{\mathscr{A}(v, u_i^{(0)})} + \mu_{P+1}\mathscr{A}(u, v), \tag{7.2}$$

we see that the eigenvalues of the operator $\tilde{T}$ defined by

$$\tilde{\mathscr{B}}(u, v) = \mathscr{A}(u, \tilde{T}v) \quad \text{for all } u \in V \tag{7.3}$$

are no smaller than those of T.

In fact,

$$\tilde{T}v = \sum_{i=1}^{P} (\mu_i^{(0)} - \mu_{P+1}^{(0)})\mathscr{A}(v, u_i^{(0)})u_i + \mu_{P+1}^{(0)}v \tag{7.4}$$

so that it has the eigenvalues $\mu_1^{(0)}, \mu_2^{(0)}, \cdots, \mu_P^{(0)}$ and then all other eigenvalues equal to $\mu_{P+1}^{(0)}$.

We see from the second monotonicity principle that the eigenvalues of $\tilde{\mathscr{B}}(v, v)/\mathscr{A}(v, v)$ on any subspace $V^{(0)}$ are still no less than those of the ratio $\mathscr{B}(v, v)/\mathscr{A}(v, v)$ on the same subspace.

If we know the first P eigenvalues $\mu_i^{(0)}$ and eigenvectors $u_i^{(0)}$ of T on the space $V_{\mathscr{A}}^{(0)}$, we form $\tilde{T}$ by (7.4) and find its eigenvalues $\tilde{\mu}_n$ on the orthogonal complement of $\phi_1, \phi_2, \cdots, \phi_k$ by the Weinstein method. These eigenvalues are then upper bounds for the eigenvalues μ_n of T on the subspace V. We now observe that

$$\tilde{T}v - \mu v = \sum_{i=1}^{P} (\mu_i^{(0)} - \mu_{P+1}^{(0)})\mathscr{A}(v, u_i^{(0)})u_i^{(0)} + (\mu_{P+1}^{(0)} - \mu)v. \tag{7.5}$$

Hence,

$$(\tilde{T} - \mu I)^{-1}w = \frac{1}{\mu_{P+1}^{(0)} - \mu}\left\{w - \sum_{i=1}^{P} \frac{\mu_i^{(0)} - \mu_{P+1}^{(0)}}{\mu_i^{(0)} - \mu}\mathscr{A}(w, u_i^{(0)})u_i^{(0)}\right\}. \tag{7.6}$$

Thus, the Weinstein determinant (4.7) becomes

$$\tilde{W}(\mu) = (\mu_{P+1}^{(0)} - \mu)^{-k} \det \left\{ \mathscr{A}(\phi_i, \phi_j) - \sum_{\alpha=1}^{P} \frac{\mu_\alpha^{(0)} - \mu_{P+1}^{(0)}}{\mu_\alpha^{(0)} - \mu} \mathscr{A}(\phi_i, u_\alpha^{(0)}) \overline{\mathscr{A}(\phi_j, u_\alpha^{(0)})} \right\}.$$

We then obtain upper bounds for the eigenvalues μ_n by means of a rational function of μ which contains only finite sums. Since $\mu_{P+1}^{(0)}$ is an eigenvalue of infinite order of $\tilde{T}$, only the roots of the equation $\tilde{W}(\mu) = 0$ which lie above $\mu_{P+1}^{(0)}$ will give upper bounds.

In order to obtain convergence of the upper bounds $\tilde{\mu}_n$ we must let P as well as k grow.

Similar remarks apply to the Aronszajn method. We can replace T_0 by the operator $\tilde{T}_0$ defined by (7.4) and still obtain upper bounds for the μ_n. If we do this, the entries of the determinant (5.12) become rational functions of μ which can be computed as finite sums. Again, only the roots above $\mu_{P+1}^{(0)}$ give upper bounds.

The fact that the rational equation $\tilde{W}(\mu) = 0$ gives upper bounds in the Weinstein method was discovered by the author (1959). Bazley and Fox (1961), (1963) realized that the process involved could be interpreted as a truncation and thereby were able to extend the idea to the Aronszajn method.

8. Error estimates for the Weinstein–Aronszajn methods. In order to bound the error in the upper bounds obtained from the Weinstein–Aronszajn method, we first prove the following lemma of Aronszajn (1948).

LEMMA (Aronszajn's inequality). *Let $R^{(1)}$ and $R^{(2)}$ be two orthogonal subspaces of the Hilbert space R with norm $\mathscr{A}(u, u)^{1/2}$ which have the property that every element v of R can be written as a sum $v = r_1 + r_2$ with r_1 in $R^{(1)}$ and r_2 in $R^{(2)}$. Let $\mathscr{B}(u, u)$ be a bounded positive semidefinite quadratic functional in V. Denote the eigenvalues of the Rayleigh quotient $\mathscr{B}(u, u)/\mathscr{A}(u, u)$ on $V_{\mathscr{A}}$ by $\nu_1 \geqq \nu_2 \geqq \cdots$. Let $\nu_1^{(1)} \geqq \nu_2^{(1)} \geqq \cdots$ and $\nu_1^{(2)} \geqq \nu_2^{(2)} \geqq \cdots$ denote the eigenvalues on the spaces $R^{(1)}$ and $R^{(2)}$, respectively. Then*

$$\nu_{l+n-1} \leqq \nu_l^{(1)} + \nu_n^{(2)}, \qquad l, n = 1, 2, \cdots. \tag{8.1}$$

Proof. Let $u_i^{(1)}$ and $u_i^{(2)}$ be the eigenfunctions in $R^{(1)}$ and $R^{(2)}$, respectively. Let R_{ln} denote the subspace of elements of R for which $\mathscr{A}(v, u_1^{(1)}) = \cdots = \mathscr{A}(v, u_{l-1}^{(1)}) = \mathscr{A}(v, u_1^{(2)}) = \cdots = \mathscr{A}(v, u_{n-1}^{(2)}) = 0$. Let $P^{(i)}$ be the orthogonal projection onto $R^{(i)}$, $i = 1, 2$. Then any v in R has the orthogonal decomposition

$$v = P^{(1)}v + P^{(2)}v.$$

If v is in R_{lm}, then

$$\mathscr{B}(P^{(1)}v, P^{(1)}v) \leqq \nu_l^{(1)} \mathscr{A}(P^{(1)}v, P^{(1)}v),$$

$$\mathscr{B}(P^{(2)}v, P^{(2)}v) \leqq \nu_n^{(2)} \mathscr{A}(P^{(2)}v, P^{(2)}v).$$

Hence by the triangle inequality and Schwarz's inequality

$$\begin{aligned}\mathscr{B}(v, v) &= \mathscr{B}(P^{(1)}v + P^{(2)}v, P^{(1)}v + P^{(2)}v)\\ &\leqq [\mathscr{B}(P^{(1)}v, P^{(1)}v)^{1/2} + \mathscr{B}(P^{(2)}v, P^{(2)}v)^{1/2}]^2\\ &\leqq [v_l^{(1)1/2}\mathscr{A}(P^{(1)}v, P^{(1)}v)^{1/2} + v_n^{(2)1/2}\mathscr{A}(P^{(2)}v, P^{(2)}v)]^2\\ &\leqq [v_l^{(1)} + v_n^{(2)}][\mathscr{A}(P^{(1)}v, P^{(1)}v) + \mathscr{A}(P^{(2)}v, P^{(2)}v)]\\ &= [v_l^{(1)} + v_n^{(2)}]\mathscr{A}(v, v)\end{aligned}$$

for all v in R_{ln}. The inequality (8.1) is now an immediate consequence of the Courant–Weyl principle.

If not all the eigenvectors needed for the definition of R_{ln} exist, we can lower one or both indices (l, n) without increasing the right-hand side of (8.1). Hence the inequality still holds and the proof is complete.

We shall apply this result to the Weinstein intermediate problem. We are given the successive maxima $\mu_1^{(0)} \geqq \mu_2^{(0)} \geqq \cdots$ of the Rayleigh quotient $\mathscr{B}(u, u)/\mathscr{A}(u, u)$ in the Hilbert space $V^{(0)}$. We assume that $\mathscr{B}(u, u)$ is positive semidefinite. In order to obtain upper bounds for the eigenvalues $\mu_1 \geqq \mu_2 \geqq \cdots$ on the subspace V, we choose $\phi_1, \phi_2, \cdots, \phi_k$ orthogonal to V. Let the orthogonal complement of $\phi_1, \cdots, \phi_k$ be the space R. The eigenvalues $\tilde{\mu}_1 \geqq \tilde{\mu}_2 \geqq \cdots$ on R are computable by means of the Weinstein determinant. Let $R^{(1)}$ be the space $V_{\mathscr{A}}$ and let $R^{(2)}$ be the subspace of elements of $V^{(0)}$ which are orthogonal to V and to $\phi_1, \phi_2, \cdots, \phi_k$.

We now apply the lemma with $n = 1$. We have $v_l = \tilde{\mu}_l$ and $v_l^{(1)} = \mu_l$ while $v_1^{(2)}$ is, by definition, the supremum of $\mathscr{B}(u, u)/\mathscr{A}(u, u)$ on $R^{(2)}$. Thus

$$\tilde{\mu}_l \leqq \mu_l + \sup_{\substack{\mathscr{A}(u,v)=0,\ \text{for all } v\in V,\\ \mathscr{A}(u,\phi_i)=0,\ i=1,2,\cdots,k}} \frac{\mathscr{B}(u, u)}{\mathscr{A}(u, u)}. \tag{8.2}$$

Since $\mu_l \leqq \tilde{\mu}_l$, the second term on the right yields a bound for the error, provided it can be bounded.

We first consider a special choice of the elements ϕ_i. Let P be the orthogonal projection onto $V_{\mathscr{A}}$, and put

$$\phi_i = u_i^{(0)} - Pu_i^{(0)}, \qquad i = 1, \cdots, k, \tag{8.3}$$

where $u_1^{(0)}, u_2^{(0)}, \cdots$ are the eigenvectors on $V^{(0)}$. Since $R^{(2)}$ is orthogonal to V,

$$\mathscr{A}(u, u_i^{(0)}) = \mathscr{A}(u, u_i^{(0)} - Pu_i^{(0)}) = \mathscr{A}(u, \phi_i) = 0, \qquad i = 1, \cdots, k,$$

for u in $R^{(2)}$. Then by definition

$$\frac{\mathscr{B}(u, u)}{\mathscr{A}(u, u)} \leqq \mu_{k+1}^{(0)}.$$

Consequently, for the special choice (8.3) we obtain the upper and lower bounds (Weinberger (1952a))

$$\tilde{\mu}_l - \mu_{k+1}^{(0)} \leqq \mu_l \leqq \tilde{\mu}_l \tag{8.4}$$

from (8.2). If $\mathscr{B}(u,u)$ is completely continuous with respect to $\mathscr{A}(u,u)$, $\mu_{k+1}^{(0)}$ approaches zero as $k \to \infty$ so that these bounds are arbitrarily close.

In practice it is seldom possible to compute the elements $u_i - Pu_i$ explicitly. We therefore examine a more general approximation.

Suppose that we can choose $\phi_1, \phi_2, \cdots, \phi_k$ in such a way that there are linear combinations $\psi_1, \psi_2, \cdots, \psi_n$ of the ϕ_i which approximate $u_1^{(0)} - Pu_1^{(0)}, \cdots, u_n^{(0)} - Pu_n^{(0)}$ in the norm $\mathscr{A}(u,u)^{1/2}$. For u in $R^{(2)}$ we have

$$\begin{aligned} |\mathscr{A}(u, u_i^{(0)})|^2 &= |\mathscr{A}(u, u_i^{(0)} - Pu_i^{(0)} - \psi_i)|^2 \\ &\leqq \mathscr{A}(u,u)\mathscr{A}(u_i^{(0)} - Pu_i^{(0)} - \psi_i, u_i^{(0)} - Pu_i^{(0)} - \psi_i). \qquad i = 1, \cdots, n. \end{aligned}$$

We write the partial Fourier series decomposition

$$u = \sum_{i=1}^{n} \frac{\mathscr{A}(u, u_i^{(0)})}{\mathscr{A}(u_i^{(0)}, u_i^{(0)})} u_i^{(0)} + w,$$

where

$$\mathscr{A}(w, u_i^{(0)}) = 0, \qquad i = 1, \cdots n.$$

Then

$$\begin{aligned} \mathscr{B}(u,u) &= \sum_{i=1}^{n} \mu_i^{(0)} \frac{|\mathscr{A}(u, u_i^{(0)})|^2}{\mathscr{A}(u_i^{(0)}, u_i^{(0)})} + \mathscr{B}(w,w) \\ &\leqq \mu_1^{(0)} \sum_{i=1}^{n} \frac{|\mathscr{A}(u, u_i^{(0)})|^2}{\mathscr{A}(u_i^{(0)}, u_i^{(0)})} + \mu_{n+1}^{(0)} \mathscr{A}(w,w) \\ &= [\mu_1^{(0)} - \mu_{n+1}^{(0)}] \sum_{i=1}^{n} \frac{|\mathscr{A}(u, u_i^{(0)})|^2}{\mathscr{A}(u_i^{(0)}, u_i^{(0)})} + \mu_{n+1}^{(0)} \mathscr{A}(u,u) \\ &\leqq \left\{ \mu_{n+1}^{(0)} + [\mu_1^{(0)} - \mu_{n+1}^{(0)}] \sum_{i=1}^{n} \frac{\mathscr{A}(u_i^{(0)} - Pu_i^{(0)} - \psi_i, u_i^{(0)} - Pu_i^{(0)} - \psi_i)}{\mathscr{A}(u_i^{(0)}, u_i^{(0)})} \right\} \\ &\quad \times \mathscr{A}(u,u). \end{aligned}$$

Thus (8.2) gives the bounds

$$\begin{aligned} (8.5) \quad \tilde{\mu}_l - \mu_{n+1}^{(0)} - [\mu_1^{(0)} - \mu_{n+1}^{(0)}] \sum \frac{\mathscr{A}(u_i^{(0)} - Pu_i^{(0)} - \psi_i, u_i^{(0)} - Pu_i^{(0)} - \psi_i)}{\mathscr{A}(u_i^{(0)}, u_i^{(0)})} \\ \leqq \mu_l \leqq \tilde{\mu}_l. \end{aligned}$$

We summarize our result in the following theorem.

THEOREM 8.1. *Let $\mathscr{B}(u,u)$ be positive semidefinite. Let there be linear combinations $\psi_1, \psi_2, \cdots, \psi_n$ of $\phi_1, \phi_2, \cdots, \phi_k$ for which the ratios $\mathscr{A}(u_i^{(0)} - Pu_i^{(0)} - \psi_i, u_i^{(0)} - Pu_i^{(0)} - \psi_i)/\mathscr{A}(u_i, u_i)$ are small. If $\tilde{\mu}_1 \geqq \tilde{\mu}_2 \geqq \cdots$ are the upper bounds computed from $\phi_1, \cdots, \phi_k$ by means of the Weinstein method, the inequalities (8.5) are valid.*

Remark 1. The inequality (8.5) is always valid. However the upper and lower bounds can be made arbitrarily close only if $\mathscr{B}(u, u)$ is completely continuous with respect to $\mathscr{A}(u, u)$ so that $\mu_{n+1}^{(0)} \to 0$ as $n \to \infty$. Truncation of the operator does not harm anything as long as $\mu_{n+1}^{(0)}$ is no smaller than the eigenvalue $\mu_{P+1}^{(0)}$ where the operator is truncated.

Remark 2. Increasing n tends to decrease $\mu_{n+1}^{(0)}$, but increases the sum on the left. A balance must be struck to obtain the best possible error estimate from given bounds.

Remark 3. When A is an elliptic operator, $u_i^{(0)} - Pu_i^{(0)}$ is often the solution of a boundary value problem. If the regularity of $u_i^{(0)} - Pu_i^{(0)}$ can be estimated a priori, bounds for the sum on the left may come, as in the Rayleigh–Ritz method, from general approximability results about the set of ϕ_i.

Example 8.1. We wish to approximate the eigenvalues of the problem

$$\begin{aligned} -\Delta u &= \lambda u \quad \text{in } D: x^2 + y^2 < 1, \\ u &= 0 \quad \text{on } \Sigma_+ : x^2 + y^2 = 1, \quad y \geqq 0, \\ \frac{\partial u}{\partial n} + u &= 0 \quad \text{on } \Sigma_- : x^2 + y^2 = 1, \quad y < 0. \end{aligned} \tag{8.6}$$

We let

$$\mathscr{A}(u, u) = \iint_D |\text{grad } u|^2 \, dx \, dy + \int_{\partial D} |u|^2 \, ds,$$

$$\mathscr{B}(u, u) = \iint_D |u|^2 \, dx \, dy.$$

The eigenvalues λ_i are the successive minima of the Rayleigh quotient $\mathscr{A}(u, u)/\mathscr{B}(u, u)$ on the space $V_{\mathscr{A}}$ which is the set of functions of $H^1(D)$ which vanish on Σ_+. We let $V^{(0)} = H^1(D)$. Then the eigenvalues $\lambda_i^{(0)}$ are eigenvalues of the problem

$$-\Delta u = \lambda u \quad \text{in } D,$$

$$\frac{\partial u}{\partial n} + u = 0 \quad \text{on } \partial D.$$

The eigenfunctions $u_n^{(0)}$ of this problem are all of the form $e^{in\theta} J_n(r\sqrt{\lambda})$ in polar coordinates. Here $J_n(r)$ is the usual Bessel function of the first kind. The eigenvalues are the roots of the equation

$$\sqrt{\lambda} J_n'(\sqrt{\lambda}) + J_n(\sqrt{\lambda}) = 0$$

for each nonnegative integer n. We obtain an infinite sequence of λ for each n, and we must order these λ in nondecreasing order. Setting $\mu_i^{(0)} = 1/\lambda_i^{(0)}$, we have $\mu_1^{(0)} \geqq \mu_2^{(0)} \geqq \cdots$. We thus know the eigenvalues and eigenfunctions for the problem on $V^{(0)}$. We do not have an explicit expression for the resolvent operator $(T - \mu I)^{-1}$ and hence we must use truncation.

According to Green's theorem we have

$$\mathscr{A}(v, \phi) = -\int_D\int v\Delta\phi\, dx\, dy + \int_{\partial D} v\left(\frac{\partial\phi}{\partial n} + \phi\right) ds.$$

Hence ϕ is orthogonal to V if and only if

$$\Delta\phi = 0 \quad \text{in } D, \tag{8.7}$$
$$\frac{\partial\phi}{\partial n} + \phi = 0 \quad \text{on } \Sigma_-.$$

For an arbitrary function u in $V^{(0)}$ we wish to have the decomposition

$$u = Pu + (u - Pu),$$

where $Pu \in V_{\mathscr{A}}$ and $(u - Pu)$ is orthogonal to V. Since Pu must vanish on Σ_+, we find that $w \equiv u - Pu$ is the solution of the mixed problem

$$\Delta w = 0 \quad \text{in } D,$$
$$w = u \quad \text{on } \Sigma_+,$$
$$\partial w/\partial n + w = 0 \quad \text{on } \Sigma_-.$$

This is not an easy problem to solve. However, if we let $\sigma(\theta)$ be the boundary value of $\partial w/\partial n + w$, we find that

$$w(r, \theta) = \int_0^{2\pi} q(r, \theta - \phi)\sigma(\phi)\, d\phi,$$

where

$$q(r, \theta) = \frac{1}{2\pi r}\left\{\cos\theta \log\frac{1}{1 + r^2 - 2r\cos\theta} + 2\sin\theta\sin^{-1}\left(\frac{r\sin\theta}{\sqrt{1 + r^2 - 2r\cos\theta}}\right) - r\right\}.$$

We must have $\sigma = 0$ on Σ_-; that is, for $\theta > \pi$. The boundary condition on Σ_+ now gives the integral equation

$$\int_0^{\pi} q(r, \theta - \phi)\sigma(\phi)\, d\phi = u(1, \theta), \qquad 0 < \theta < \pi.$$

This is an integral equation of the first kind and it is not easy to solve. However, if we find an approximate solution $\tilde{\sigma}(\theta)$, the function

$$\psi(r, \theta) = \int_0^{\pi} q(r, \theta - \phi)\tilde{\sigma}(\phi)\, d\phi$$

satisfies $\Delta\psi = 0$ and $\partial\psi/\partial n + \psi = 0$ on Σ_-. Hence it is orthogonal to V and may be used in the Weinstein method. In order to make ψ approximate $u - Pu$ in norm, it is sufficient to make the integral of $|\tilde{\sigma} - \sigma|^2$ small.

We recall that we derived the Aronszajn method as a special case of the Weinstein method. We can therefore apply Theorem 8.1 to the Aronszajn method if the quadratic functional $\mathscr{D}(\{u, v, w\}, \{u, v, w\})$ is positive semidefinite and if we can find the orthogonal projection P from H_0 to H.

We see immediately from the definition (5.14) that $\mathscr{D}$ is positive semidefinite if and only if:

(a) $\mathscr{B}_0(u, u)$ is positive semidefinite.

(b) $\mathscr{B}'(u, u)$ is zero for all u.

That is, N must be the zero operator, and

$$\mathscr{B}(u, u) = \mathscr{B}_0(u, u)$$

is positive semidefinite.

We now seek to find the projection operator P onto H.

We recall that H_0 consists of equivalence classes of triples $\{u, v, w\}$ of elements of V. The subspace H consists of elements of the form $\{v, v, v\}$. Its orthogonal complement consists of elements of the form $\{-(\kappa I - T_0)^{-1}(M\zeta + N\psi), \kappa^{-1}\zeta, \psi\}$ where ζ and ψ are arbitrary elements of $V_{\mathscr{A}_0}$. The eigenvectors of $\mathscr{S}$ on H_0 which correspond to positive eigenvalues are of the form $\{u_i^{(0)}, 0, 0\}$, where $u_i^{(0)}$ is an eigenvector of T_0. To find $P\{u_i^{(0)}, 0, 0\}$, we must have a decomposition

$$\{u_i^{(0)}, 0, 0\} = \{v, v, v\} + \{-(\kappa I - T_0)^{-1}(M\zeta + N\psi), \kappa^{-1}\zeta, \psi\}. \tag{8.8}$$

Then the first element is $P\{u_i^{(0)}, 0, 0\}$ and the second element is $\{u_i^{(0)}, 0, 0\} - P\{u_i^{(0)}, 0, 0\}$, which we wish to approximate.

Since $\mathscr{B}'(u, u) = 0$ for all u, an element $\{u, v, w\}$ is equivalent to $\{u, v, w'\}$ for any w'. Hence we may let $\psi = 0$ and write equations for the first two components of (8.8). The first gives

$$v - (\kappa I - T_0)^{-1}M\zeta = u_i^{(0)}. \tag{8.9}$$

The second component gives the equation

$$v + \kappa^{-1}\zeta = 0.$$

We solve for ζ and substitute in (8.9).

Thus we find that v must satisfy the equation

$$v + \kappa(\kappa I - T_0)^{-1}Mv = u_i^{(0)}.$$

Operating with $(\kappa I - T_0)$, we find

$$-T_0 v + \kappa v + \kappa M v = (\kappa - \mu_i^{(0)})u_i^{(0)}.$$

If we can solve this equation, we have the projection $P\{u_i^{(0)}, 0, 0\} = \{v, v, v\}$. Then up to equivalence

$$\{u_i^{(0)},0,0\} - P\{u_i^{(0)},0,0\} = \{-(\kappa I - T_0)^{-1}M\zeta_i^*, \kappa^{-1}\zeta_i^*, 0\},$$

where we have let $\zeta_i^* = -\kappa v$. That is, ζ_i^* is the solution of the problem

$$T_0\zeta_i^* - \kappa\zeta_i^* - \kappa M\zeta_i^* = \kappa(\kappa - \mu_i^{(0)})u_i^{(0)}. \tag{8.10}$$

We recall that the successive maxima of the Rayleigh quotients are

$$\nu_i = \mu_i/(\kappa - \mu_i), \qquad \tilde{\nu}_i = \mu_i/(\kappa - \tilde{\mu}_i), \quad \text{and} \quad \nu_i^{(0)} = \mu_i^{(0)}/(\kappa - \mu_i^{(0)}).$$

The inequality (8.4) now shows that if we apply the Aronszajn method using the trial elements $\zeta_1^*, \cdots, \zeta_r^*$, then

$$\frac{\tilde{\mu}_l}{\kappa - \tilde{\mu}_l} - \frac{\mu_{r+1}^{(0)}}{\kappa - \mu_{r+1}^{(0)}} \leqq \frac{\mu_l}{\kappa - \mu_l} \leqq \frac{\tilde{\mu}_l}{\kappa - \tilde{\mu}_l},$$

or equivalently,

$$\tilde{\mu}_l - \frac{(\kappa - \tilde{\mu}_l)^2\mu_{r+1}^{(0)}}{\kappa^2 - 2\mu_{r+1}^{(0)}\kappa + \tilde{\mu}_l\mu_{r+1}^{(0)}} \leqq \mu_l \leqq \tilde{\mu}_l.$$

Since $\tilde{\mu}_l$, $\mu_{r+1}^{(0)}$, and μ_l do not depend on κ and since κ is only constrained to be above $\mu_1^{(0)}$, we may let κ approach $\mu_1^{(0)}$ to obtain the error bounds

$$\tilde{\mu}_l - \frac{(\mu_1^{(0)} - \tilde{\mu}_l)^2\mu_{r+1}^{(0)}}{\mu_1^{(0)2} - 2\mu_{r+1}^{(0)}\mu_1^{(0)} + \tilde{\mu}_l\mu_{r+1}^{(0)}} \leqq \mu_l \leqq \tilde{\mu}_l, \qquad l = 1, 2, \cdots. \tag{8.11}$$

If we cannot find the ζ_i^* exactly, we approximate them in the norm of H_0 by elements ζ_i'. We note that by definition (5.13),

$$\mathscr{C}(\{-(\kappa I - T_0)^{-1}M\zeta, \kappa^{-1}\zeta, \psi\}, \{-(\kappa I - T_0)^{-1}M\zeta, \kappa^{-1}\zeta, \psi\})^{1/2}$$
$$= [\mathscr{A}_0((\kappa I - T_0)^{-1}M\zeta, M\zeta) + \kappa^{-1}\mathscr{A}'(\zeta, \zeta)]^{1/2}.$$

It is in this norm that the ζ_i' must approximate the ζ_i^*. On the other hand,

$$\mathscr{C}(\{u_i^{(0)},0,0\},\{u_i^{(0)},0,0\}) = (\kappa - \mu_i^{(0)})\mathscr{A}_0(u_i^{(0)}, u_i^{(0)}).$$

Theorem 8.1 applied to this situation now gives the following theorem.

THEOREM 8.2. *Let $\mathscr{B}_0(u,u) = \mathscr{B}(u,u)$ be positive semidefinite. Let $\mathscr{A}_0(u,u)$ be positive definite, let*

$$\mathscr{A}'(u,u) = \mathscr{A}_0(u, Mu)$$

be positive semidefinite, and let $\mathscr{A}(u,u) = \mathscr{A}_0(u,u) + \mathscr{A}'(u,u)$. Let

$$\mathscr{B}(u,u) = \mathscr{A}_0(T_0u, u)$$

be bounded by a multiple of $\mathscr{A}_0(u,u)$. Let the eigenvalues $\mu_i^{(0)}$ and eigenvectors $\mu_i^{(0)}$ of the operator T_0 as well as the resolvent $(T_0 - \mu I)^{-1}$ be known. The Aronszajn method is applied to find upper bounds $\tilde{\mu}_i$ for the eigenvalues μ_i of the problem $T_0u = \mu(u + Mu)$, using trial functions $\zeta_1, \cdots, \zeta_a$. If there exist linear combinations

$\zeta_1', \cdots, \zeta_k'$ *of these* ζ_i *which approximate the solutions* ζ_i^* *of* (8.10) *in the sense that*

$$\frac{\kappa(\mu_1^{(0)} - \mu_{k+1}^{(0)})}{\kappa - \mu_1^{(0)}} \times \sum_{i=1}^{k} \frac{\mathscr{A}_0((\kappa I - T_0)^{-1}(M\zeta_i' - M\zeta_i^*), M\zeta_i' - M\zeta_i^*) + \kappa^{-1}\mathscr{A}'(\zeta_i' - \zeta_i^*, \zeta_i' - \zeta_i^*)}{(\kappa - \mu_i^{(0)})\mathscr{A}_0(u_i^{(0)}, u_i^{(0)})} \leqq \varepsilon^2 \tag{8.12}$$

with some $\kappa > \mu_1^{(0)}$ *and*

$$\varepsilon^2 < \frac{\kappa^2 - 2\mu_{k+1}^{(0)}\kappa + \tilde{\mu}_l\mu_{k+1}^{(0)}}{\kappa - \tilde{\mu}_l},$$

then

$$\tilde{\mu}_l - \frac{(\kappa - \tilde{\mu}_l)^2[\mu_{k+1}^{(0)} + \varepsilon^2]}{\kappa^2 - 2\mu_{k+1}^{(0)}\kappa + \mu_{k+1}^{(0)}\tilde{\mu}_l - \varepsilon^2(\kappa - \tilde{\mu}_l)} \leqq \mu_l \leqq \tilde{\mu}_l. \tag{8.13}$$

9. Complementary bounds in terms of matrix eigenvalues. We observe that upper bounds for the eigenvalues μ_n cannot be obtained from just the set of computed numbers $\mathscr{A}(v_i, v_j)$ and $\mathscr{B}(v_i, v_j)$ for some finite set $v_1, \cdots, v_k$ if V is of dimension greater than k. For if Pu is the projection of u orthogonal to the v_i, these numbers are the same for the forms $\mathscr{B}(u, u)$ and

$$\mathscr{B}'(u, u) \equiv \mathscr{B}(u, u) + c\mathscr{A}(Pu, Pu)$$

for any constant c. If w is any element of V that is orthogonal to $v_1, \cdots, v_k$, we see that

$$\mu_1' = \sup_{u \in V} \frac{\mathscr{B}'(u, u)}{\mathscr{A}(u, u)} \geqq \frac{\mathscr{B}(w, w)}{\mathscr{A}(w, w)} + c,$$

which may be arbitrarily large.

In the Weinstein and Aronszajn methods the needed additional information consists of the eigenvalues and eigenvectors of a related problem. In the error bounds of the Rayleigh–Ritz method it consists of the a priori bounds on the derivatives of the eigenfunctions which are needed to establish approximability results like (2.1).

We shall now show how upper bounds can be computed from a matrix eigenvalue problem when a finite set of a priori data is available. The results presented here and in the next section have appeared only in the author's technical report (1959) and lecture notes (1962).

We first suppose that we are given elements $p_1, p_2, \cdots, p_m$ of $V_{\mathscr{A}}$ and a real number σ with the property

$$\begin{aligned} &\mathscr{A}(v, p_1) = \mathscr{A}(v, p_2) = \cdots = \mathscr{A}(v, p_m) = 0, \quad v \in V \\ &\text{implies} \quad \mathscr{B}(v, v) \leqq \sigma\mathscr{A}(v, v). \end{aligned} \tag{9.1}$$

We wish to find upper bounds for the eigenvalues of the Rayleigh quotient $\mathscr{B}(v, v)/\mathscr{A}(v, v)$ on the completion $V_{\mathscr{A}}$ of V.

We choose elements $v_1, v_2, \cdots, v_k$ of $V_{\mathscr{A}}$, and define the $(m + k) \times (m + k)$ matrices

$$F_k = \left(\begin{array}{c|c} \mathscr{A}(p_i, p_j) & \mathscr{A}(p_i, (\sigma I - T)v_\beta) \\ \hline \mathscr{A}((\sigma I - T)v_\alpha, p_j) & \mathscr{A}((\sigma I - T)v_\alpha, (\sigma I - T)v_\beta) \end{array}\right) \tag{9.2}$$

and

$$G_k = \left(\begin{array}{c|c} 0 & -\mathscr{A}(p_i, v_\beta) \\ \hline -\mathscr{A}(v_\alpha, p_j) & -\mathscr{A}((\sigma I - T)v_\alpha, v_\beta) \end{array}\right). \tag{9.3}$$

The operator T is, as usual, defined by the identity $\mathscr{B}(u, v) = \mathscr{A}(u, Tv)$ for all u, v in $V_{\mathscr{A}}$.

The following theorem gives a procedure for finding upper bounds for those eigenvalues μ_i which lie above σ.

THEOREM 9.1. *Let the matrix $\mathscr{A}(p_i, v_\alpha)$ be of rank m and let G_k be nonsingular. Let $\tau_1 \geqq \tau_2 \geqq \cdots$ be the m largest eigenvalues of the matrix problem*

$$F_k\mathbf{z} = \tau G_k\mathbf{z}. \tag{9.4}$$

If $\tilde{\mu}_i$ is defined by

$$\tilde{\mu}_i = \sigma + \tau_i, \tag{9.5}$$

then

$$\mu_i \leqq \tilde{\mu}_i, \qquad i = 1, 2, \cdots, m.$$

Proof. Let Q be any linear operator on $V_{\mathscr{A}}$ whose range is the subspace of linear combinations of the v_α and for which

$$\mathscr{A}(u - Qu, p_i) = 0 \quad \text{for } i = 1, \cdots, m \quad \text{for all } u \in V_{\mathscr{A}}. \tag{9.6}$$

Because of (9.1) the sesquilinear form

$$\tilde{\mathscr{B}}(u, v) \equiv \mathscr{B}(u, v) + \sigma\mathscr{A}(u - Qu, v - Qv) - \mathscr{B}(u - Qu, v - Qv)$$

satisfies

$$\tilde{\mathscr{B}}(u, u) \geqq \mathscr{B}(u, u).$$

Hence the eigenvalues $\tilde{\mu}_1 \geqq \tilde{\mu}_2 \geqq \cdots$ of the Rayleigh quotient $\tilde{\mathscr{B}}(u, u)/\mathscr{A}(u, u)$ satisfy

$$\tilde{\mu}_i \geqq \mu_i, \qquad i = 1, 2, \cdots.$$

Moreover, if we define the operator Q^* by

$$\mathscr{A}(u, Qv) = \mathscr{A}(Q^*u, v) \quad \text{for all } u, v \text{ in } V,$$

we can write

$$\tilde{\mathscr{B}}(u, v) = \mathscr{A}(u, \sigma v - Q^*(\sigma I - T)(I - Q)v - (\sigma I - T)Qv).$$

Thus, $\tilde{\mathscr{B}}(u, v) = \mathscr{B}(u, \tilde{T}v)$ where

$$\tilde{T} = \sigma I - [Q^*(\sigma I - T)(I - Q) + (\sigma I - T)Q].$$

Since the inverse of $\sigma I - \mu I$ is just $(\sigma - \mu)^{-1}I$ and since the operator in brackets has finite range, we can find the eigenvalues $\tilde{\mu}_i$ by the methods of § 3.

If $k > m$, the operator Q is not uniquely determined by (9.6). In order to obtain the best possible bounds, we define Qu as that linear combination of the v_α which minimizes $\sigma\mathscr{A}(u - Qu, u - Qu) - \mathscr{B}(u - Qu, u - Qu)$ under the constraint (9.6). A standard argument with Lagrange multipliers shows that Qu is to satisfy the conditions

$$\mathscr{A}\left(v_\alpha, (\sigma I - T)Qu + \sum_{i=1}^{m} a_i p_i\right) = \mathscr{A}(v_\alpha, (\sigma I - T)u), \quad \alpha = 1, \cdots, k \tag{9.7}$$

for some constants a_i.

In order to compute Q, we observe that the eigenvalues τ_i of (9.4) remain unchanged if we replace the p_i and the v_α by linearly independent linear combinations. We choose these combinations so that

$$\mathscr{A}(v_\nu, p_i) = 0 \quad \text{for } i = 1, \cdots, m \quad \text{and} \quad \nu > m. \tag{9.8}$$

Since G_k is nonsingular, (9.1) shows that the matrix $\mathscr{A}((\sigma I - T)v_\mu, v_\nu)$ with $\mu, \nu > m$ is positive definite. Therefore we can make

$$\mathscr{A}((\sigma I - T)v_\mu, v_\nu) = \delta_{\mu\nu} \quad \text{for } \mu, \nu > m. \tag{9.9}$$

We now choose $v_1, \cdots, v_m$ so that

$$\mathscr{A}((\sigma I - T)v_\mu, v_i) = 0 \quad \text{for } \mu > m, \quad i \leqq m, \tag{9.10}$$

and then choose $p_1, \cdots, p_m$ so that

$$\mathscr{A}(v_i, p_j) = \delta_{ij} \quad \text{for } i, j = 1, \cdots, m. \tag{9.11}$$

Then (9.6) and (9.7) imply

$$Qu = \sum_{i=1}^{m} \mathscr{A}(u, p_i)v_i + \sum_{\nu=m+1}^{k} \mathscr{A}(u, (\sigma I - T)v_\nu)v_\nu,$$

so that

$$\begin{aligned}\tilde{T}u = \sigma u - \Bigg[&\sum_{i=1}^{m} \mathscr{A}\left(u, \left\{(\sigma I - T)v_i - \sum_{j=1}^{m} \mathscr{A}((\sigma I - T)v_i, v_j)p_j\right\}\right)p_i \\ &+ \sum_{i=1}^{m} \mathscr{A}(u, p_i)(\sigma I - T)v_i + \sum_{\nu=m+1}^{k} \mathscr{A}(u, (\sigma I - T)v_\nu)(\sigma I - T)v_\nu\Bigg].\end{aligned}$$

We can now apply Theorem 3.1 with $(u, v) \equiv \mathscr{A}(u, v)$, $A_0 = I$, $B_0 = \sigma I$, $C = 0$, and D the operator in brackets. Since $(B_0 - \mu A_0)^{-1} = (\sigma - \mu)^{-1}I$, the matrix (3.4) becomes

$$
\left(\begin{array}{c|c|c}
\begin{array}{l}\delta_{ij} - (\sigma - \mu)^{-1} \\ \times \Big[\mathscr{A}(p_j, (\sigma I - T)v_i \\ \quad - \sum_{l=1}^{m} \mathscr{A}(p_j, p_l) \\ \quad \times \mathscr{A}(v_l, (\sigma I - T)v_i)\Big]\end{array} & \begin{array}{l}-(\sigma - \mu)^{-1} \\ \times \Big[\mathscr{A}((\sigma I - T)v_s, (\sigma I - T)v_i) \\ \quad - \sum_{l=1}^{m} \mathscr{A}((\sigma I - T)v_s, p_l) \\ \quad \times \mathscr{A}(v_l, (\sigma I - T)v_i)\Big]\end{array} & \begin{array}{l}-(\sigma - \mu)^{-1} \\ \times \Big[\mathscr{A}((\sigma I - T)v_\nu, (\sigma I - T)v_i) \\ \quad - \sum_{l=1}^{m} \mathscr{A}((\sigma I - T)v_\nu, p_l) \\ \quad \times \mathscr{A}(v_l, (\sigma I - T)v_i)\Big]\end{array} \\
\hline
-(\sigma - \mu)^{-1}\mathscr{A}(p_j, p_r) & \delta_{rs} - (\sigma - \mu)^{-1}\mathscr{A}((\sigma I - T)v_s, p_r) & -(\sigma - \mu)^{-1}\mathscr{A}((\sigma I - T)v_\nu, p_r) \\
\hline
-(\sigma - \mu)^{-1}\mathscr{A}(p_j, (\sigma I - T)v_\kappa) & \begin{array}{l}-(\sigma - \mu)^{-1}\mathscr{A}((\sigma I - T)v_s, \\ \qquad (\sigma I - T)v_\kappa)\end{array} & \begin{array}{l}\delta_{\kappa\nu} - (\sigma - \mu)^{-1}\mathscr{A}((\sigma I - T)v_\nu, \\ \qquad (\sigma I - T)v_\kappa)\end{array}
\end{array}\right)
$$

$$i, j, r, s = 1, \cdots, m, \quad \kappa, \nu = m + 1, \cdots, k.$$

We add $\mathscr{A}(v_r, (\sigma I - T)v_i)$ times the matrix in the middle rows to the matrix in the upper rows. We then interchange the first m rows with the middle m rows, multiply all elements by $\mu - \sigma$, and replace δ_{ij} and $\delta_{\kappa\nu}$ by means of (9.9) and (9.11). In this way we obtain the matrix $F_k - (\mu - \sigma)G_k$.

Since the above operations do not alter the rank of the matrix when $\mu \neq \sigma$, we see from Theorem 3.1 that $\mu \neq \sigma$ is an eigenvalue of $\tilde{T}$ if and only if $\tau = \mu - \sigma$ is an eigenvalue of (9.4) and that, moreover, the multiplicities of these eigenvalues are the same. Thus the theorem is verified for the positive eigenvalues τ. If there are l such positive eigenvalues, the upper bounds $\tilde{\mu}_i$ for $i > l$ are either σ or $\sigma + \tau_j$ for some $j \leqq i$. We shall now show that $\tau_m \geqq 0$. It will then follow that $\tau_j = 0$ for $l < j \leqq m$ and hence that $\tilde{\mu}_i = \sigma + \tau_i$ for $i \leqq m$.

We write $\mathbf{z}$ in terms of its components as

$$\mathbf{z} \sim \begin{pmatrix} a_1 \\ \vdots \\ a_m \\ b_1 \\ \vdots \\ b_k \end{pmatrix},$$

and we define the elements $p = \sum_{i=1}^{m} \bar{a}_i p_i$ and $v = \sum_{\alpha=1}^{k} \bar{b}_\alpha v_\alpha$.

Then

$$\mathbf{z} \cdot F_k \mathbf{z} = \mathscr{A}(p + (\sigma I - T)v, p + (\sigma I - T)v).$$

Hence the matrix F_k is positive semidefinite.

If there were a complex eigenvalue τ of (9.4), we would have

$$\mathbf{z} \cdot F_k \mathbf{z} = \tau \mathbf{z} \cdot G_k \mathbf{z}.$$

Taking imaginary parts, we see that then $\mathbf{z} \cdot G_k\mathbf{z} = \mathbf{z} \cdot F_k\mathbf{z} = 0$. Since F_k is positive semidefinite, this implies $F_k\mathbf{z} = 0$ and hence also $G_k\mathbf{z} = 0$. Since G_k is nonsingular, it follows that $\mathbf{z} = 0$. That is, the problem (9.4) has only real eigenvalues.

We now observe that on the m-dimensional space of $\mathbf{z}$ with the b_i equal to zero, $\mathbf{z} \cdot G_k\mathbf{z}$ is zero. On the k-dimensional space where $b_1 = \cdots = b_m = 0, \mathbf{z} \cdot G\mathbf{z}$ is negative definite. It follows from the Remark after Theorem 5.1, Chap. 3, that G_k has k negative and m positive eigenvalues. The same is then true of $\tau G_k - F_k$ for sufficiently large positive τ. On the other hand, at $\tau = 0$ this matrix is negative semidefinite. It follows by continuity that the m highest eigenvalues of $\tau G_k - F_k$ go from nonpositive values to positive values as τ goes from zero to infinity. Therefore the problem (9.4) has m nonnegative eigenvalues $\tau_1 \geqq \tau_2 \geqq \cdots \geqq \tau_m \geqq 0$, and these give upper bounds according to (9.5).

Remark 1. If F_k is positive definite (that is, if the elements $p_1, \cdots, p_m, (\sigma I - T)v_1, \cdots, (\sigma I - T)v_k$ are linearly independent), $\tau = 0$ is not an eigenvalue of (9.4). It is then computationally advantageous to find the positive eigenvalues $\kappa_1 \leqq \kappa_2 \leqq \cdots \leqq \kappa_m$ of the eigenvalue problem

$$G_k\mathbf{z} = \kappa F_k\mathbf{z}$$

and set $\tau_i = 1/\kappa_i$ in (9.5).

Remark 2. Since the quadratic form of the matrix $G_k - \tau^{-1}F_k$ is nondecreasing in τ, its eigenvalues can only change from negative to positive values as τ increases. The same is then true of $\tau G_k - F_k$ for $\tau > 0$. It follows that τ_i is that value of τ for which the $(m + 1 - i)$th of the eigenvalues in descending order of the matrix $\tau G_k - F_k$ is zero.

Remark 3. It follows from Remark 2 that the upper bound $\tilde{\mu}_i$ is nonincreasing in k (that is, when more v_α are used) but nondecreasing in m (when it takes more p_i to obtain the condition (9.1)).

Remark 4. If rank $[\mathscr{A}(p_i, v_\alpha)] =$ rank $[G_k] - k = r < m$, we choose the p_i so that $p_{r+1}, p_{r+2}, \cdots, p_m$ are orthogonal to all the v_α. We can then apply Theorem 9.1 to obtain upper bounds for the first r eigenvalues $\mu_1^{(m-r)} \geqq \mu_2^{(m-r)} \geqq \cdots \geqq \mu_r^{(m-r)}$ of the Rayleigh quotient $\mathscr{B}(u, u)/\mathscr{A}(u, u)$ on the orthogonal complement $V^{(m-r)}$ of $p_{r+1}, \cdots, p_m$. If we denote the first r eigenvalues of (9.4) by $\tau_{m-r+1} \geqq \cdots \geqq \tau_m$, we see from the separation theorem (Theorem 9.1, Chap. 3) that $\mu_{i+m-r} \leqq \mu_i^{(m-r)} \leqq \sigma + \tau_{i+m-r}$. Thus

$$\mu_i \leqq \sigma + \tau_i, \qquad i = m - r + 1, \cdots, m. \tag{9.12}$$

Theorem 9.1 is then still valid if we make the convention that the first $m - r$ eigenvalues of (9.4) are $+\infty$. Moreover, Remarks 2 and 3 are still valid. G_k has r positive eigenvalues and an $(m - r)$-dimensional null space. Note that we may have $k < m$.

Remark 5. The explicit appearance of the operator T, which is formally $A^{-1}B$, may pose a computational problem. If B^{-1} is easily found, it is easier to choose $w_1, \cdots, w_k$ in $V_{\mathscr{A}}$ and then to determine $v_\alpha = AB^{-1}w_\alpha$ so that $w_\alpha = Tv_\alpha$. Note, however, that the w_α must be chosen so that the $AB^{-1}v_\alpha$ are again in $V_{\mathscr{A}}$.

Remark 6. In order to investigate the convergence of the bounds in Theorem 9.1 we let $\mu_1 \geqq \mu_2 \geqq \cdots \geqq \mu_s > \sigma \geqq \mu_{s+1}$ and let $u_1, u_2, \cdots$ be the eigenvectors of T.

If $s < m$, we arrange the p_i so that $\mathscr{A}(p_i, u_\alpha) = 0$ for $\alpha \leqq s, i > s$ and that $\mathscr{A}(p_i, (\sigma I - T)^{-1}p_j) = 0$ for $i \leqq s, j > s$. (We raise σ slightly if necessary so that it is in the resolvent set.) Since a nontrivial linear combination of $u_\alpha, \alpha \leqq s$ orthogonal to the p_i would violate (9.1), we can make $\mathscr{A}(p_i, u_\alpha) = \delta_{i\alpha}$ for $i \leqq s$. Since $p_{s+1}, \cdots, p_m$ are orthogonal to $u_1, \cdots, u_s$, the same is true of $(\sigma I - T)^{-1}p_i$. Therefore the matrix $\mathscr{A}((\sigma I - T)^{-1}p_i, p_j)$ with $i, j \geqq s + 1$ is positive definite, and we can rearrange the p_i so that $\mathscr{A}((\sigma I - T)^{-1}p_i, p_j) = \delta_{ij}$ for $i, j = s + 1, \cdots, m$. If we choose $k = m$ and

$$v_\alpha = \begin{cases} u_\alpha & \alpha \leqq s, \\ (\sigma I - T)^{-1}p_\alpha, & s + 1 \leqq \alpha \leqq m, \end{cases}$$

then the rank of $\mathscr{A}(p_i, v_\alpha)$ is m. Moreover,

$$Qu = \sum_{i=1}^{s} \mathscr{A}(u, p_i)u_i + \sum_{i=s+1}^{m} \mathscr{A}(u, p_i)(\sigma I - T)^{-1}p_i.$$

In particular, $Qu_i = u_i$ for $i \leqq s$. It follows that $\tilde{T}u_i = \mu_i u_i$ for $i \leqq s$. For u orthogonal to $u_1, \cdots, u_s$ we see that

$$\begin{aligned}\tilde{\mathscr{B}}(u, u) &= \mathscr{B}(u, u) + \mathscr{A}((\sigma I - T)(u - Qu), u - Qu) \\ &= \sigma\mathscr{A}(u, u) - \sum_{i=s+1}^{m} |\mathscr{A}(u, p_i)|^2 + \sum_{i=1}^{s} (\sigma - \mu_i)|\mathscr{A}(u, p_i)|^2.\end{aligned}$$

Since $\mu_i > \sigma$ for $i \leqq s, \tilde{\mathscr{B}}(u, u) \leqq \sigma\mathscr{A}(u, u)$ for u orthogonal to $u_1, \cdots, u_s$. Thus $\tilde{\mu}_i = \mu_i$ for $i \leqq s$, and $\tilde{\mu}_{s+1} = \sigma$. A similar but simpler proof gives the same result if $s = m$.

We see from Remark 3 and continuity, that in general *if* $\mu_i > \sigma$ *and if there are linear combinations of the* v_α *which are close in norm to* $u_1, \cdots, u_i, (\sigma I - T)^{-1}p_1, \cdots, (\sigma I - T)^{-1}p_m$, *then the bound* $\tilde{\mu}_i$ *is close to* μ_i. In particular, this proves convergence without the hypothesis of complete continuity, which is needed to prove the convergence of the Weinstein and Aronszajn methods.

If, as in the Aronszajn method, we know the first $m + 1$ eigenvalues $\mu_1^{(0)} \geqq \mu_2^{(0)} \geqq \cdots \geqq \mu_{m+1}^{(0)}$ and the eigenvectors $u_1^{(0)}, \cdots, u_m^{(0)}$ of the Rayleigh quotient $\mathscr{B}_0(u, u)/\mathscr{A}(u, u)$ where $\mathscr{B}_0(u, u) \geqq \mathscr{B}(u, u)$, we have (9.1) with $p_i = u_i^{(0)}$ and $\sigma = \mu_{m+1}^{(0)}$. If we choose the first m v_α to be the $u_i^{(0)}$, we are led to the bounds obtained from the Aronszajn method applied to the truncated operator of § 7. However, the determination of the zeros of a determinant whose entries are polynomials has been replaced by the simpler computation needed to find the eigenvalues of the problem (9.4).

If $\mu_1^{(0)} \geqq \cdots \geqq \mu_{m+1}^{(0)} > 0$ are the first $m + 1$ eigenvalues and $u_1^{(0)}, \cdots, u_m^{(0)}$ are the first m eigenvectors of the Rayleigh quotient $\mathscr{B}_0(u, u)/\mathscr{A}_0(u, u)$ with $\mathscr{B}_0(u, u) \geqq \mathscr{B}(u, u)$ and $\mathscr{A}_0(u, u) \leqq \mathscr{A}(u, u)$ and if we know an operator $\tilde{M}$ with the property $\mathscr{A}_0(u, v) = \mathscr{A}(u, \tilde{M}v)$ for all u, v in V, then we have (9.1) with $p_i = \tilde{M}u_i^{(0)}$ and $\sigma = \mu_{m+1}^{(0)}$.

It frequently happens that the quadratic functionals $\mathscr{A}(u, u)$ and $\mathscr{B}(u, u)$ can be extended so that they satisfy the Hypotheses 1.1, Chap. 3, on a space $V^{(0)}$ containing V, and that one is given the conditions (9.1) with p_i in $V^{(0)}$ but not in V. We shall suppose that we know the operator T for which $\mathscr{A}(u, Tv) = \mathscr{B}(u, v)$ for all u and v in $V^{(0)}$.

Theorem 9.1 now gives upper bounds for the eigenvalues $\mu_i^{(0)}$ of T on $V^{(0)}$. These are, in general, bigger than the μ_i, so that we cannot obtain arbitrarily close bounds.

We note that the conditions (9.1) are not disturbed if an element which is orthogonal to V is added to each p_i.

We choose $\phi_1, \cdots, \phi_l$ in $V^{(0)}$ so that

$$\mathscr{A}(v, \phi_\nu) = 0 \qquad \text{for all } v \in V,$$

$$\mathscr{A}(\phi_\nu, \phi_\mu) = \delta_{\nu\mu} \quad \text{for } \nu, \mu = 1, \cdots, l.$$

We define the projection

$$P^{(l)}u \equiv u - \sum_{\nu=1}^{l} \mathscr{A}(u, \phi_\nu)\phi_\nu, \tag{9.13}$$

as in § 4. We may then replace each p_i by $P^{(l)}p_i$. Moreover, as we saw in § 4, the nonzero eigenvalues of the operator $P^{(l)}TP^{(l)}$ on $V^{(0)}$ are upper bounds for the eigenvalues of T on V. Thus, we may obtain upper bounds for the μ_i by applying Theorem 9.1 and Remark 4 to $P^{(l)}TP^{(l)}$ with the p_i replaced by $P^{(l)}p_i$. We thus obtain the matrices

$$F_k^{(l)} \equiv \left(\begin{array}{c|c} \mathscr{A}(P^{(l)}p_i, p_j) & \mathscr{A}(P^{(l)}p_i, (\sigma I - T)P^{(l)}v_\beta) \\ \hline \mathscr{A}((\sigma I - T)P^{(l)}v_\alpha, P^{(l)}p_j) & \mathscr{A}((\sigma I - T)P^{(l)}v_\alpha, P^{(l)}(\sigma I - T)P^{(l)}v_\beta) \end{array}\right)$$

and

$$G_k^{(l)} \equiv \begin{pmatrix} 0 & -\mathscr{A}(P^{(l)}p_i, v_\beta) \\ -\mathscr{A}(v_\alpha, P^{(l)}p_j) & -\mathscr{A}((\sigma I - T)P^{(l)}v_\alpha, P^{(l)}v_\beta) \end{pmatrix}.$$

THEOREM 9.2. *Let $\mathscr{A}(u, u)$ and $\mathscr{B}(u, u)$ satisfy Hypotheses* 1.1, *Chap.* 3, *on a space $V^{(0)}$ which contains V, and suppose that one has $p_1, \cdots, p_m$ in $V^{(0)}$ with the property* (9.1). *Let the projection operator $P^{(l)}$ be defined by* (9.13) *in terms of the orthonormal elements $\phi_1, \cdots, \phi_l$ of the orthogonal complement of V. Choose any $v_1, v_2, \cdots, v_k$ in $V^{(0)}$, and let* rank $[\mathscr{A}(v_\nu, P^{(l)}p_i)] =$ rank $[G_k] - k = r$. *Then if $\tau_{m-r+1} \geqq \cdots \geqq \tau_m$ are the r largest eigenvalues of the problem*

$$F_k^{(l)}\mathbf{z} = \tau G_k^{(l)}\mathbf{z},$$

we have the bounds

$$\mu_i \leqq \sigma + \tau_i \equiv \tilde{\mu}_i, \qquad i = m - r + 1, \cdots, m,$$

for the eigenvalues of the Rayleigh quotient $\mathscr{B}(u, u)/\mathscr{A}(u, u)$ on V.

Remark 1. If, as in the Weinstein method, the first m eigenvalues $\mu_1^{(0)} \geqq \mu_2^{(0)} \geqq \cdots \geqq \mu_{m+1}^{(0)}$ and the eigenvectors $u_1^{(0)}, \cdots, u_m^{(0)}$ of some Rayleigh quotient $\mathscr{B}_0(u, u)/\mathscr{A}(u, u)$ on $V^{(0)}$ with $\mathscr{B}_0(u, u) = \mathscr{B}(u, u)$ for all u in V are known, then (9.1) holds with $p_i = u_i^{(0)}$ and $\sigma = \mu_{m+1}^{(0)}$. The particular choice $k = m$ and $v_i = u_i^{(0)}$

leads to the bounds given by the Weinstein method applied to the truncated operator of § 7.

Remark 2. An argument like that in Remark 6 after Theorem 9.1 shows that if $\mu_i > \sigma$ and if the elements $u_1, \cdots, u_i, (\sigma I - T)^{-1}p_1, \cdots, (\sigma I - T)^{-1}p_m$ can be approximated in norm by linear combinations of the ϕ_ν and the v_α, then $\tilde{\mu}_i$ is close to μ_i. Again one need not assume complete continuity to prove convergence.

Example 9.1. Consider the problem

$$-\Delta u = \lambda u \quad \text{in } D, \qquad u = 0 \quad \text{on } \partial D, \tag{9.14}$$

in a bounded two-dimensional domain D. Suppose that the closure of D is contained in the rectangle $R: 0 < x < a,\ 0 < y < b$. The eigenvalues of the corresponding problem on R are easily seen to be

$$\pi^2\left(\frac{p^2}{a^2} + \frac{q^2}{b^2}\right), \qquad p, q = 1, 2, \cdots. \tag{9.15}$$

The corresponding eigenfunctions are

$$\sin\frac{p\pi x}{a}\sin\frac{q\pi y}{b}.$$

We order the eigenvalues (9.15) in nondecreasing order $\lambda_1^{(0)} \leqq \lambda_2^{(0)} \leqq \cdots$ and call the corresponding eigenfunctions $u_1^{(0)}, u_2^{(0)}, \cdots$.

We let $V^{(0)}$ be the space of real functions defined on R having piecewise continuous derivatives and vanishing on the boundary of R. Let V be the subspace of those functions which are zero outside D.

We define

$$\mathscr{A}(u, v) = \int_R \operatorname{grad} u \cdot \operatorname{grad} v \, dx \quad \text{for } u, v \in V^{(0)},$$

$$\mathscr{B}(u, v) = \int_D uv \, dx \quad \text{for } u, v \in V,$$

$$\mathscr{B}_0(u, v) = \int_R uv \, dx.$$

Since $\mathscr{B}_0(u, u) = \mathscr{B}(u, u)$ for u in V and since the $\lambda^{(0)}$ are the eigenvalues of the Rayleigh quotient $\mathscr{A}(u, u)/\mathscr{B}_0(u, u)$ on $V^{(0)}$, we see that (9.1) holds with $p_i = u_i^{(0)}$ and $\sigma = 1/\lambda_{m+1}^{(0)}$.

We define $\mu_i = 1/\lambda_i$ where λ_i is the ith eigenvalue of the problem (9.14).

It is easily verified by integration by parts that ϕ is orthogonal to V if and only if $\Delta\phi = 0$ in D. Since we are free to extend $\mathscr{B}(u, u)$ from V to $V^{(0)}$ in any manner we wish, the operator T is only partly defined. An integration by parts and the definition of $V^{(0)}$ show that T must only satisfy

$$-\Delta T v = v \quad \text{in } D,$$

$$T v = 0 \quad \text{on } \partial R,$$

for all v in V. For example, if $\eta(\mathbf{x})$ is any smooth function which is identically one on D and which vanishes on the boundary of R, we may take

$$(Tu)(\mathbf{x}) = \frac{-1}{2\pi}\int_R \eta(\mathbf{x}) \log|\mathbf{x} - \mathbf{y}|\eta(\mathbf{y})u(\mathbf{y})\,dy$$

for all u in $V^{(0)}$.

We can now choose any $v_1, \cdots, v_k$ in V and $\phi_1, \cdots, \phi_l$ with $\Delta\phi_\nu = 0$ in D and $\mathscr{A}(\phi_\nu, \phi_\mu) = \delta_{\nu\mu}$ and form the matrices $F_k^{(l)}$ and $G_k^{(l)}$.

Instead of constructing an explicit operator T we may proceed as in Remark 5 after Theorem 9.1. That is, we choose any $w_1, \cdots, w_k$ so that $w_\alpha = \Delta w_\alpha = 0$ on ∂R. We then put $v_\alpha = -\Delta w_\alpha$ and $Tv_\alpha = w_\alpha$.

In order to avoid the difficulty of computing the operator T altogether, we now consider the case in which $\mathscr{B}(u, u)$ is positive definite. We make the following assumptions in place of Hypotheses 1.1, Chap. 3.

HYPOTHESES 9.1. (a) *$\mathscr{B}(u, u)$ is defined and positive definite on a space $V^{(0)}$ which contains V.*

(b) *The Rayleigh quotient $\mathscr{A}(u, u)/\mathscr{B}(u, u)$ with u in V is bounded below by some constant γ.*

(c) *There is a known linear transformation L from V to $V^{(0)}$ with the property that*

$$\mathscr{A}(u, v) = \mathscr{B}(u, Lv) \quad \textit{for all } u, v \in V.$$

(Note that formally $L = B^{-1}A$.)

We then define the eigenvalues $\lambda_1 \leqq \lambda_2 \leqq \cdots$ as the successive minima of the Rayleigh quotient $\mathscr{A}(u, u)/\mathscr{B}(u, u)$ on the completion of V with respect to the norm $\{\mathscr{A}(u, u) + (1 - \gamma)\mathscr{B}(u, u)\}^{1/2}$.

We now suppose that instead of (9.1) we are given elements $p_1, \cdots, p_m$ of $V^{(0)}$ and a constant ρ with the property that

$$\begin{aligned} &\mathscr{B}(v, p_1) = \cdots = \mathscr{B}(v, p_m) = 0, \quad v \in V \\ &\text{implies } \mathscr{A}(v, v) \geqq \rho\mathscr{B}(v, v). \end{aligned} \tag{9.16}$$

If we choose any elements $v_1, \cdots, v_k$ in V, we see that because of (9.16),

$$\mathscr{A}(u, u) \geqq \mathscr{A}(u, u) - [\mathscr{A}(u - Qu, u - Qu) - \rho\mathscr{B}(u - Qu, u - Qu)],$$

where Q is any linear operator on V whose range consists of linear combinations of the v_α and which has the property that $\mathscr{B}(u - Qu, p_1) = \cdots = \mathscr{B}(u - Qu, p_m) = 0$ for all u in V. Proceeding as in the proof of Theorem 9.1, we obtain lower bounds for the λ_i in terms of the eigenvalues of the matrix problem

$$F\mathbf{z} = \xi G\mathbf{z}, \tag{9.17}$$

where now

$$F = \begin{pmatrix} \mathscr{B}(p_i, p_j) & \mathscr{B}(p_i, (L - \rho I)v_\beta) \\ \mathscr{B}((L - \rho I)v_\alpha, p_j) & \mathscr{B}((L - \rho I)v_\alpha, (L - \rho I)v_\beta) \end{pmatrix}, \tag{9.18}$$

$$G = \begin{pmatrix} 0 & -\mathscr{B}(p_i, v_\beta) \\ -\mathscr{B}(v_\alpha, p_j) & -\mathscr{B}((L - \rho I)v_\alpha, v_\beta) \end{pmatrix}. \tag{9.19}$$

(While L may be an unbounded transformation, we apply Theorem 3.1 on $V^{(0)}$ with $A_0 = I$, $B_0 = \rho I$, and L is used only in defining the fixed elements Lv_α of $V^{(0)}$.)

THEOREM 9.3. *Let Hypotheses* 9.1 *and the conditions* (9.16) *be satisfied. Choose any elements* $v_1, \cdots, v_k$ *of* V, *and let* $\operatorname{rank}[\mathscr{B}(p_i, v_\alpha)] = \operatorname{rank}[G] - k = r$. *If* $\xi_{m-r+1} \geqq \xi_{m-r+2} \geqq \cdots \geqq \xi_m$ *are the r highest eigenvalues of the problem* (9.17) *and if we define*

$$\tilde{\lambda}_i = \rho - \xi_i, \qquad i = m - r + 1, \cdots, m,$$

then

$$\lambda_i \geqq \tilde{\lambda}_i, \qquad i = m - r + 1, \cdots, m.$$

Remark 1. As in Theorem 9.1, we see that if $p_1, \cdots, p_m$, $(L - \rho I)v_1, \cdots, (L - \rho I)v_k$ are linearly independent so that F is positive definite, it is computationally advantageous to find the positive eigenvalues $\kappa_{m-r+1} \leqq \kappa_{m-r+2} \leqq \cdots \leqq \kappa_m$ of the problem

$$G\mathbf{z} = \kappa F \mathbf{z} \tag{9.20}$$

and then set $\tilde{\lambda}_i = \rho - (1/\kappa_i)$, $i = m - r + 1, \cdots, m$.

Remark 2. If V is not dense in $V^{(0)}$ in the sense of the norm $\mathscr{B}(u, u)^{1/2}$, we can also derive the analogue of Theorem 9.2.

Remark 3. If $\lambda_i < \rho$ and if there are linear combinations of the v_α which approximate $(L - \rho I)^{-1}p_1, \cdots, (L - \rho I)^{-1}p_m, u_1, \cdots, u_i$ in the sense of the norm

$$[\mathscr{B}(u, u) + \mathscr{B}(Lu, Lu)]^{1/2}, \tag{9.21}$$

then we can show as in Remark 6 after Theorem 9.1 that $\tilde{\lambda}_i$ is close to λ_i.

Example 9.2. We again consider the problem (9.14) of Example 9.1. We define $\lambda_i^{(0)}$ as in Example 9.1, but we now let $V^{(0)}$ be the set of all continuously differentiable functions on the closure of D, with

$$\mathscr{A}(u, u) = \int_D |\operatorname{grad} u|^2 \, dx,$$

$$\mathscr{B}(u, u) = \int_D u^2 \, dx.$$

We let V be the set of functions in $V^{(0)}$ which vanish on the boundary of D. We see as in Example 1 that (9.16) holds with $p_i = u_i^{(0)}$ and $\rho = \lambda_{m+1}^{(0)}$. By Theorem 9.3 we can now choose any $v_1, \cdots, v_k$ in V and form the matrices F and G with $L = -\Delta$ to obtain lower bounds $\tilde{\lambda}_i$ for the λ_i.

Because of the presence of the terms $\mathscr{B}((L - \rho I)v_\alpha, (L - \rho I)v_\beta)$ in F, we cannot take v_α in the closure $V_{\mathscr{B}}$ or even $V_{\mathscr{A}}$, but only in the closure of V with respect to the norm (9.21).

According to Theorem 3.3 the lower bounds $\tilde{\lambda}_i$ are the roots, with multiple roots repeated, of the determinantal equation

$$\det [F - (\rho - \lambda)G] = 0, \tag{9.22}$$

the lowest root being $\tilde{\lambda}_{m-r+1}$. Suppose for the moment that $\mathscr{B}(p_i, p_j) = \delta_{ij}$. We subtract the sum of $\mathscr{B}((L - \lambda I)v_\alpha, p_i)$ times the ith row of $F - (\rho - \lambda)G$ from the $(m + \alpha)$th row to find that

$$\det [F - (\rho - \lambda)G]$$
$$= \det \left[\mathscr{B}((L - \rho I)v_\alpha, (L - \lambda I)v_\beta) - \sum_{i=1}^{m} \mathscr{B}((L - \lambda I)v_\alpha, p_i)\mathscr{B}(p_i, (L - \lambda I)v_\beta) \right].$$

More generally, we define P to be the operator of orthogonal projection in the sense of the scalar product $\mathscr{B}(u, v)$ onto the set of linear combinations of the p_i. Then the determinant in (9.22) is seen to be equal to $\det \mathscr{B}(p_i, p_j)$ times

$$\det [\mathscr{B}((L - \lambda I)v_\alpha - (\rho - \lambda)v_\beta + (I - P)(L - \lambda I)v_\beta)]. \tag{9.23}$$

Thus the lower bounds $\tilde{\lambda}_{m-r+1} \leqq \tilde{\lambda}_{m-r+2} \leqq \cdots \leqq \tilde{\lambda}_m$ are the first r zeros (with multiple zeros repeated) of this $k \times k$ determinant, whose elements are quadratic rather than linear in λ.

If k is small it is more convenient to obtain the zeros of (9.23) directly than to find the eigenvalues of the problem (9.17).

Example 9.3. Consider the problem

$$-\frac{d}{dx}\left[(1 + x^2)\frac{du}{dx}\right] = \lambda u \quad \text{for } 0 < x < 1, \tag{9.24}$$
$$u(0) = 0,$$
$$u'(1) = 0.$$

We let

$$\mathscr{B}(u, u) = \int_0^1 u^2\, dx,$$
$$\mathscr{A}(u, u) = \int_0^1 (1 + x^2)u'^2\, dx = \int_0^1 uLu\, dx,$$

where

$$Lu \equiv -\frac{d}{dx}\left[(1 + x^2)\frac{du}{dx}\right].$$

V is the set of smooth functions which satisfy the boundary conditions.

We let

$$\mathscr{A}_0(u, u) = \int_0^1 u'^2\, dx. \tag{9.25}$$

The eigenvalues $\lambda_1^{(0)} \leqq \lambda_2^{(0)} \leqq \cdots$ and eigenfunctions $u_n^{(0)}$ of the problem for the Rayleigh quotient $\mathscr{A}_0(u, u)/\mathscr{B}(u, u)$ are those of the problem

$$-d^2u/dx^2 = \lambda u,$$
$$u(0) = 0, \tag{9.26}$$
$$u'(1) = 0.$$

Thus, $\lambda_n^{(0)} = (n - \frac{1}{2})^2\pi^2$, $u_n^{(0)} = \sin(n - \frac{1}{2})\pi x$, $n = 1, 2, \cdots$. We now choose $p_i = u_i^{(0)}, i = 1, \cdots, m$. If u is in V and $\mathscr{B}(u, p_1) = \cdots = \mathscr{B}(u, p_m) = 0$, then

$$\mathscr{A}(u, u) \geqq \mathscr{A}_0(u, u) \geqq (m + \tfrac{1}{2})^2\pi^2\mathscr{B}(u, u).$$

Thus we have (9.16) with $\rho = (m + \frac{1}{2})^2\pi^2$.

Choose, for instance, $m = 3$ and

$$v_n = p_n = \sin(n - \tfrac{1}{2})\pi x, \qquad n = 1, 2, 3.$$

Substituting in (9.23), we obtain a 3×3 determinant instead of the determinant of the 6×6 matrix $F - (\rho - \lambda)G$. We compute the lower bounds $\tilde{\lambda}_i$. In order to have an idea of the degree of approximation, we also compute Rayleigh–Ritz upper bounds λ_i' using the same v_α. We find

$$\tilde{\lambda}_1 = 2.67 \leqq \lambda_1 \leqq 2.73,$$
$$\tilde{\lambda}_2 = 27.85 \leqq \lambda_2 \leqq 28.19, \tag{9.27}$$
$$\tilde{\lambda}_3 = 77.03 \leqq \lambda_3 \leqq 82.19.$$

Thus the bounds with just three trial functions give reasonable approximations to the eigenvalues.

We observe that if we change the boundary conditions in (9.24) to

$$u(0) = 0,$$
$$u'(1) + u(1) = 0, \tag{9.28}$$

we have

$$\mathscr{A}(u, u) = \int_0^1 (1 + x^2)u'^2\,dx + u(1)^2.$$

If $\mathscr{A}_0$ is given by (9.25), we still have $\mathscr{A}(u, u) \geqq \mathscr{A}_0(u, u)$. The minimum problem for $\mathscr{A}_0(u, u)/\mathscr{B}(u, u)$ on the space V of functions which satisfy the boundary conditions (9.28) still leads to the problem (9.26). Thus, if we let

$$p_n = \sin(n - \tfrac{1}{2})\pi x, \qquad n = 1, \cdots, m,$$

we again have (9.16) with $\rho = (m + \frac{1}{2})^2\pi^2$. However, the p_n do not satisfy the boundary conditions (9.28) and hence we must use different v_α.

10. Simpler upper and lower bounds. In this section we present some simple upper and lower bounds for the eigenvalues $\lambda_1 \leqq \lambda_2 \leqq \cdots$ of the problem

$$Au = \lambda Bu, \qquad u \in V,$$

where both (u, Au) and (u, Bu) are positive definite on V. We base our results on Theorem 9.3.

We recall that if the condition (9.16) is satisfied and if the rank of $\mathscr{B}(p_i, v_\alpha)$ is r and that of G is $r + k$, then the r lowest zeros $\tilde{\lambda}_{m-r+1} \leqq \tilde{\lambda}_{m-r+2} \leqq \cdots \leqq \tilde{\lambda}_m$ of the determinant (9.23) (with multiple zeros repeated) are lower bounds for the generalized eigenvalues $\lambda_{m-r+1} \leqq \lambda_{m-r+2} \leqq \cdots \leqq \lambda_m$ of $Au = \lambda Bu$. Those $\tilde{\lambda}_i$ which lie below ρ are just those values of $\lambda < \rho$ for which the matrix

$$(10.1) \quad -\mathscr{B}((L - \lambda I)v_\alpha, v_\beta) + (\rho - \lambda)^{-1}\mathscr{B}((I - P)(L - \lambda I)v_\alpha, (I - P)(L - \lambda I)v_\beta)$$

has the eigenvalue zero. The multiplicity of the eigenvalue zero at λ is equal to the multiplicity of λ as a zero of (9.23). (We recall that P is the orthogonal projection in the sense of the scalar product $\mathscr{B}(u, v)$ onto the space of linear combinations of the p_i.)

If we divide (10.1) by $-\lambda$ and take the limit as $\lambda \to -\infty$, we obtain the matrix

$$(10.2) \qquad -\mathscr{B}(Pv_\alpha, Pv_\beta).$$

This matrix is certainly negative semidefinite. Its rank is the rank r of the matrix $\mathscr{B}(v_\alpha, p_i)$. Thus, this matrix has r negative eigenvalues and the eigenvalue zero with multiplicity $k - r$.

The derivative with respect to λ of the quadratic form of (10.1) with $v = \sum x_\alpha v_\alpha$ is

$$(10.3) \quad \mathscr{B}\left(\frac{1}{\rho - \lambda}(I - P)(L - \lambda I)v - v, \frac{1}{\rho - \lambda}(I - P)(L - \lambda I)v - v\right).$$

If we take v in the null space of the matrix (10.2) (that is, if $Pv = 0$), we see from (10.3) that the derivative of the quadratic form of (10.1) is

$$(\rho - \lambda)^{-2}\mathscr{B}((I - P)(L - \rho I)v, (I - P)(L - \rho I)v).$$

If this were zero, we would have

$$(I - P)(L - \rho I)v = 0.$$

Writing $v = \sum x_\alpha v_\alpha$, $-P(L - pI)v = \sum a_i p_i$, and $\mathbf{z} \sim \{\bar{a}_1, \cdots, \bar{a}_m, \bar{x}_1, \cdots, \bar{x}_k\}$, we then have $G\mathbf{z} = 0$. Since the rank deficiency of G is equal to that of $\mathscr{B}(p_i, v_\alpha)$, this implies that $x_1 = \cdots = x_k = 0$ and hence that $v = 0$. Thus if v lies in the null space of the limiting matrix $\mathscr{B}(Pv_\alpha, Pv_\beta)$ and is not zero, then the quadratic form of (10.1) is strictly increasing.

It follows that for sufficiently negative λ the matrix (10.1) has r negative eigenvalues and $k - r$ positive eigenvalues with respect to the matrix $\mathscr{B}(v_\alpha, v_\beta)$. We write these eigenvalues as

$$\kappa_1(\lambda) \geqq \kappa_2(\lambda) \geqq \cdots \geqq \kappa_k(\lambda).$$

Since the derivative of the quadratic form (10.1) is (10.3) which is nonnegative, we see from the second monotonicity principle that each $\kappa_i(\lambda)$ is nondecreasing in λ and only the last r become negative.

We can then characterize the lower bound $\tilde{\lambda}_i$ as that value of λ for which $\kappa_{k-m+i}(\lambda)$ changes from being negative to being positive. Thus

$$\kappa_{k-m+i}(\lambda) > 0 \quad \text{for } \tilde{\lambda}_i < \lambda < \rho. \tag{10.4}$$

We now apply the inequality (9.5), Chap. 3, to the matrix (10.1). The eigenvalues of the matrix $-\mathscr{B}((L-\lambda I)v_\alpha, v_\beta)$ with respect to $\mathscr{B}(v_\alpha, v_\beta)$ are $\lambda - \lambda_1' \geqq \lambda - \lambda_2' \geqq \cdots \geqq \lambda - \lambda_k'$, where the λ_i' are the upper bounds for the eigenvalues λ_i which are obtained by applying the Rayleigh–Ritz method with the trial functions $v_1, \cdots, v_k$.

If we denote the eigenvalues of the matrix

$$\mathscr{B}((I-P)(L-\lambda I)v_\alpha, (I-P)(L-\lambda I)v_\beta) \tag{10.5}$$

by $\kappa_1^{(2)}(\lambda) \geqq \kappa_2^{(2)}(\lambda) \geqq \cdots \geqq \kappa_k^{(2)}(\lambda)$, the inequality (9.5), Chap. 3, states that

$$\kappa_{k-m+i}(\lambda) \leqq \lambda - \lambda_i' + \frac{1}{\rho - \lambda}\kappa_{k-m+1}^{(2)}(\lambda). \tag{10.6}$$

Let $\phi(\lambda)$ be some upper bound for $\kappa_{k-m+1}^{(2)}(\lambda)$. (Note that $\kappa_{k-m+1}^{(2)}$ is the mth of the eigenvalues of the matrix (10.3) arranged in nondecreasing order. It does not increase if more elements v_α are added.) Then if $\tilde{\lambda}_i < \lambda < \rho$, we have from (10.4) and (10.6),

$$0 < \kappa_{k-m+i}(\lambda) \leqq \lambda - \lambda_i' + \frac{\phi(\lambda)}{\rho - \lambda}.$$

Therefore, if we find any value $\lambda_i^* < \rho$ for which

$$\lambda_i^* - \lambda_i' + \frac{\phi(\lambda_i^*)}{\rho - \lambda_i^*} \leqq 0, \tag{10.7}$$

λ_i^* must lie below $\tilde{\lambda}_i$ and hence also below λ_i. We have established the following method for obtaining both upper and lower bounds from the Rayleigh–Ritz method.

THEOREM 10.1. *Let the Hypotheses* 9.1 *of this chapter and* 1.1 *of Chap.* 3 *be fulfilled. Suppose we are given elements* $p_1, \cdots, p_m$ *and a* $\rho > 0$ *such that* (9.16) *holds. Let* $\lambda_1' \leqq \lambda_2' \leqq \cdots \leqq \lambda_k'$ *be the eigenvalues of the matrix* $\mathscr{B}(Lv_\alpha, v_\beta)$ *with respect to* $\mathscr{B}(v_\alpha, v_\beta)$ *where the* v_α *are in* V. *Let* $\phi(\lambda)$ *be an upper bound for the m-th of the eigenvalues of the matrix* (10.5) *with respect to* $\mathscr{B}(v_\alpha, v_\beta)$, *arranged in ascending order.* (*P is the orthogonal projection in the scalar product* $\mathscr{B}(u, v)$ *onto the space of linear combinations of the* p_i.) *Then if* λ_i^* *is any number below* ρ *for which* (10.7) *holds, we have the upper and lower bounds*

$$\lambda_i^* \leqq \lambda_i \leqq \lambda_i'. \tag{10.8}$$

Remark 1. Computationally, one usually prefers to solve the equation

$$\lambda_i^* - \lambda_i' + \frac{\phi(\lambda_i^*)}{\rho - \lambda_i^*} = 0.$$

However, the inequality (10.7) shows how one may err in approximating the solution of this equation and still get a lower bound.

Remark 2. Theorem 10.1 allows us to compute lower bounds for the λ_i from the Rayleigh–Ritz eigenvalues without the need to compute the lower bounds $\tilde{\lambda}_i$. In the same way, we may compute an upper bound from $\tilde{\lambda}_i$ without computing λ_i' by simply letting $\lambda = \tilde{\lambda}_i$ in (10.6) and recalling that $\kappa_{k-m+1}(\tilde{\lambda}_i) = 0$.

We then find that

$$\lambda_i' \leqq \tilde{\lambda}_i + \frac{\phi(\tilde{\lambda}_i)}{\rho - \tilde{\lambda}_i}$$

so that we obtain the bounds

$$\tilde{\lambda}_i \leqq \lambda_i \leqq \tilde{\lambda}_i + \frac{\phi(\tilde{\lambda}_i)}{\rho - \tilde{\lambda}_i} \tag{10.9}$$

in terms of the $\tilde{\lambda}_i$ alone.

The above error bounds are simpler when the p_i lie in V and we choose $k = m$ and $v_i = p_i$. In this case $(I - P)v_i = 0$ so that the matrix (10.1) and hence also $\phi(\lambda)$ is independent of λ. This case has also been treated by Knauer (1971).

A further simplification occurs if there is a transformation $\hat{L}$ from V to $V^{(0)}$ with the property that its first $m + 1$ eigenvalues $\hat{\lambda}_1 \leqq \hat{\lambda}_2 \leqq \cdots \leqq \hat{\lambda}_{m+1}$ and the eigenvectors $p_1, \cdots, p_m$ are known, and that the difference $\hat{L} - L$ is bounded in the sense that

$$\mathscr{B}(\hat{L}u - Lu, \hat{L}u - Lu) \leqq a\mathscr{B}(u, u) \tag{10.10}$$

for all u in V. In this case we see that

$$\mathscr{B}(u, p_i) = 0, \qquad i = 1, \cdots, m,$$

implies that

$$\mathscr{B}(\hat{L}u, u) \equiv \hat{\mathscr{A}}(u, u) \geqq \hat{\lambda}_{m+1}\mathscr{B}(u, u).$$

Since

$$|\mathscr{B}(\hat{L}u - Lu, u)|^2 \leqq \mathscr{B}(u, u)\mathscr{B}(\hat{L}u - Lu, \hat{L}u - Lu) \leqq a\mathscr{B}(u, u)^2,$$

we see that $\mathscr{B}(u, p_i) = 0$ implies

$$\mathscr{B}(Lu, u) \geqq (\hat{\lambda}_{m+1} - \sqrt{a})\mathscr{B}(u, u).$$

Thus we have (9.16) with $\rho = \hat{\lambda}_{m+1} - \sqrt{a}$.

If we now choose $k = m$ and $v_i = p_i$, we see that $(I - P)(\hat{L} - \lambda)v_i = 0$. Therefore the matrix (10.5) becomes

$$\mathscr{B}((L - \hat{L})p_i, (I - P)(L - \hat{L})p_j).$$

It follows from (10.10) that we may take $\phi(\lambda) = a$. By (10.7), we may put $\lambda_i^* = \frac{1}{2}[\lambda_i' + \rho - \{(\lambda_i' - \rho)^2 + 4a\}^{1/2}]$. Thus if the trial functions $p_1, \cdots, p_m$ are used in the Rayleigh–Ritz method, we find the bounds (see Koehler (1957),

Börsch-Supan (1958), Bertram (1957), Weinberger (1954))

$$(10.11)\quad \lambda_i' - \tfrac{1}{2}[\{(\hat{\lambda}_{m+1} - \sqrt{a} - \lambda_i')^2 + 4a\}^{1/2} - (\hat{\lambda}_{m+1} - \sqrt{a} - \lambda_i')] \leqq \lambda_i \leqq \lambda_i'.$$

The well-known inequality $\sqrt{1 + \varepsilon^2} \leqq 1 + \frac{1}{2}\varepsilon^2$ gives the weaker but simpler bounds

$$(10.12)\quad \lambda_i' - a(\hat{\lambda}_{m+1} - \sqrt{a} - \lambda_i')^{-1} \leqq \lambda_i \leqq \lambda_i'.$$

Thus if $\hat{\lambda}_{m+1}$ can be made arbitrarily large by choosing m large, we find that the difference between the upper and lower bounds in (10.11) can be made arbitrarily small. If we replace λ_i' by any upper bound, say $\hat{\lambda}_i + \sqrt{a}$, we obtain an a priori bound for the error in the Rayleigh–Ritz method with the special choice of functions $p_1, p_2, \cdots, p_m$.

Example 10.1. Consider the problem

$$-\Delta u + f(x, y)u = \lambda u \quad \text{in } D\colon\ 0 < x < 1, \quad 0 < y < 1,$$

$$u = 0 \quad \text{on } \partial D.$$

We take

$$\mathscr{B}(u, v) = \int_D\int uv\,dx\,dy,$$

and

$$Lu = -\Delta u + f(x, y)u.$$

Let

$$\hat{L}u = -\Delta u.$$

Suppose that

$$|f(x, y)|^2 \leqq a.$$

Then (10.10) is satisfied. The eigenfunctions of $\hat{L}$ on V are just

$$p_{rs} = \sin r\pi x \sin s\pi y.$$

If we use all the p_{rs} with $\pi^2(r^2 + s^2) \leqq \sigma^2$ in the Rayleigh–Ritz method, we obtain the bounds (10.11) or (10.12) with $\hat{\lambda}_{m+1}$ replaced by σ^2. This improves a bound of Hammerstein (1927).

11. Inclusion theorems. If we replace the projection operator P in the matrix (10.1) by the zero operator, the quadratic form of this matrix for each λ is not decreased. By the second monotonicity principle the same is true of its eigenvalues. That is, for $\lambda < \rho$ the eigenvalues of the matrix

$$(11.1)\quad -\mathscr{B}((L - \lambda I)v_\alpha, v_\beta) + \frac{1}{\rho - \lambda}\mathscr{B}((L - \lambda I)v_\alpha, (L - \lambda I)v_\beta)$$

do not lie below those of the matrix (10.1). We recall that the lower bound $\tilde{\lambda}_i$ was characterized by the fact that the eigenvalue $\kappa_{k-m+i}(\lambda)$ of the matrix (10.1) is positive for $\lambda > \tilde{\lambda}_i$ and negative for $\lambda < \tilde{\lambda}_i$. Let $\tilde{\tilde{\lambda}}_i$ be the value of λ (if any) at which the $(k - m + i)$th of the eigenvalues of (11.1) arranged in descending order is zero. Then by what we have said, $\tilde{\tilde{\lambda}}_i \leqq \tilde{\lambda}_i \leqq \lambda_i$, so that $\tilde{\tilde{\lambda}}_i$ is also a lower bound for λ_i.

We note that $(\rho - \lambda)$ times (11.1) is equal to

$$\mathscr{B}((L - \lambda I)v_\alpha, (L - \rho I)v_\beta). \tag{11.2}$$

We assume that we are not so fortunate that a linear combination of the v_α is an eigenvector of L. Then the matrix (11.2) at $\lambda = \rho$ is positive definite. Hence the eigenvalues of (11.1) approach infinity as $\lambda \to \rho$. Moreover, a differentiation shows that the quadratic form, and hence the eigenvalues, are increasing in λ. The zero of the determinant of (11.1) immediately below ρ must be $\tilde{\tilde{\lambda}}_m$, the one below that $\tilde{\tilde{\lambda}}_{m-1}$, and so forth.

It is, of course, easier to find the zeros of the determinant of (11.2) which is just (11.1) multiplied by $\rho - \lambda$.

The matrix (11.1) does not depend on the p_i. Thus ρ may be any lower bound for λ_{m+1}, and we do not need to know the p_i. We pay for our ignorance of the p_i by obtaining the worse lower bounds $\tilde{\tilde{\lambda}}_i$ instead of $\tilde{\lambda}_i$.

We have thus obtained the following result of Lehmann (1949), (1963) and Maehly (1952).

THEOREM 11.1 (Lehmann and Maehly). *Let ρ be any lower bound for λ_{m+1}. Let the roots of the equation*

$$\det \mathscr{B}((L - \lambda I)v_\alpha, (L - \rho I)v_\beta) = 0 \tag{11.3}$$

below ρ with multiple roots repeated be $\gamma_1 \leqq \gamma_2 \leqq \cdots \leqq \gamma_s < \rho$. Then

$$\lambda_{m-s+j} \geqq \gamma_j, \qquad j = 1, \cdots, s.$$

Remark. Instead of the determinantal equation (11.3) we can write the eigenvalue problem

$$-\sum_{\beta=1}^{k} \mathscr{B}((L - \rho I)v_\alpha, v_\beta)x_\beta = \frac{1}{\rho - \lambda} \sum_{\beta=1}^{k} \mathscr{B}(L - \rho I)v_\alpha, (L - \rho I)v_\beta)x_\beta.$$

Example 11.1. If in the problem (9.24) we use the same $v_i = p_i$ as before but compute the upper bounds from (11.3) instead of (9.23), we obtain the lower bounds $\tilde{\tilde{\lambda}}_1 = 2.52$ instead of $\tilde{\lambda}_1 = 2.67$; $\tilde{\tilde{\lambda}}_2 = 27.8$ instead of $\tilde{\lambda}_2 = 27.84$; $\tilde{\tilde{\lambda}}_3 = 75.2$ instead of $\tilde{\lambda}_3 = 77.03$. Thus the bounds are slightly worse, but they involve less labor.

If it is known that $\lambda_n \to \infty$ as $n \to \infty$, any number ρ is a lower bound for some λ_{m+1}. If $\rho > \lambda_1$, then there is an m such $\lambda_m < \rho \leqq \lambda_{m+1}$. If $\rho < \lambda_1$, we see from Theorem 11.1 that (11.3) cannot have any roots below ρ. We can thus restate Theorem 11.1 in the following form.

COROLLARY 1. *If $\lambda_n \to \infty$ as $n \to \infty$ and if for some ρ the equation* (11.3) *has roots $\gamma_1 \leqq \gamma_2 \leqq \cdots \leqq \gamma_s < \rho$, then there are at least $s - j + 1$ eigenvalues λ_i in the interval*

$$\gamma_j \leqq \lambda < \rho.$$

Such a result is called an *inclusion theorem*. The simplest case occurs when there is only one trial function ($k = 1$). The equation (11.3) becomes

$$\mathscr{B}((L - \lambda I)v_1, (L - \rho I)v_1) = 0 \tag{11.4}$$

which is easily solved for λ. If $\lambda < \rho$, we see from Corollary 1 that there is an eigenvalue between λ and ρ. If $\lambda > \rho$, we simply interchange the roles of ρ and λ to reach the same conclusion. If $\lambda = \rho$, we see that ρ is an eigenvalue. Thus there is always an eigenvalue on the closed interval determined by ρ and λ. This inclusion theorem was found by Kohn (1947). (See also Temple (1928), Temple and Bickley (1933), Kato (1949).)

We see that the values λ and ρ occur symmetrically in (11.4). If we change the variables to

$$\rho = t + a, \qquad \lambda = t - a,$$

(11.4) becomes

$$\mathscr{B}(Lv_1, Lv_1) - 2t\mathscr{B}(Lv_1, v_1) + (t^2 - a^2)\mathscr{B}(v_1, v_1) = 0.$$

We can assign any t and solve for a to obtain a symmetric interval $|\lambda - t| \leqq a$ which contains an eigenvalue. A reasonable choice of t is one which minimizes the length $2a$ of the interval. This leads to

$$t = \frac{\mathscr{B}(Lv_1, v_1)}{\mathscr{B}(v_1, v_1)}, \qquad a = \left\{\frac{\mathscr{B}(Lv_1, Lv_1)}{\mathscr{B}(v_1, v_1)} - \frac{|\mathscr{B}(Lv_1, v_1)|^2}{\mathscr{B}(v_1, v_1)^2}\right\}^{1/2}.$$

This form of the inclusion theorem is due to Krylov and Bogoliubov (1929), but it was formulated more explicitly by D. H. Weinstein (1934). An interesting mechanism for generating inclusion theorems was found by Washizu (1952), (1955).

A better inclusion theorem for λ_1 in terms of a lower bound for λ_2 was given by Trefftz (1933).

Various other inclusion theorems for matrices were given by Collatz (1942), Walker and Weston (1949), Wielandt (1953), (1954), Kreyszig (1954), (1955), Nitsche (1959) and Kahan (1967).

An inclusion theorem in which the trial function does not have to satisfy boundary conditions was found by Fox, Henrici, and Moler (1967) and improved by Moler and Payne (1968).

12. The method of G. Fichera. Trefftz (1933) discovered the following result about a real symmetric positive semidefinite completely continuous integral operator on an N-dimensional domain D. Let

$$(Tu)(x) \equiv \int_D K(x, y)u(y)\,dy \tag{12.1}$$

with $K(x, y) = K(y, x)$. If T has the eigenvalues μ_i and eigenfunctions u_i with $\int_D u_i u_j\,dx = \delta_{ij}$, then $K(x, y)$ has the formal expansion

$$K(x, y) \sim \sum_{i=1}^{\infty} \mu_i u_i(x) u_i(y).$$

If, moreover, the integral of $K(x, y)^2$ is finite, then

$$\int_D \int_D K(x, y)^2 \, dx \, dy = \sum_{i=1}^{\infty} \mu_i^2.$$

Suppose lower bounds $\mu_1' \geqq \mu_2' \geqq \cdots \geqq \mu_k'$ for $\mu_1, \mu_2, \cdots, \mu_k$ are computed by the Rayleigh–Ritz method. Then if $n \leqq k$, one finds

$$\begin{aligned}\int_D \int_D K(x, y)^2 \, dx \, dy &= \sum_{i=1}^{\infty} \mu_i^2 \\ &\geqq \sum_{i=1}^{n-1} \mu_i'^2 + \mu_n^2 + \sum_{i=n+1}^{k} \mu_i'^2 \\ &= \sum_{i=1}^{k} \mu_i'^2 + \mu_n^2 - \mu_n'^2.\end{aligned}$$

Thus we have the Trefftz upper bound

$$\mu_n^2 \leqq \mu_n'^2 + \int_D \int_D K(x, y)^2 \, dx \, dy - \sum_{i=1}^{k} \mu_i'^2. \tag{12.2}$$

De Vito (1961) noted that if one lets

$$K^{(m)}(x, y) = \int_{D} \cdots \int_D K(x, x_1) K(x_1, x_2) \cdots K(x_{m-1}, y) \, dx_1 \cdots dx_{m-1}$$

be the kernel corresponding to T^m, then

$$\int_D K^{(m)}(x, x) \, dx = \sum_{i=1}^{\infty} \mu_i^m,$$

and that if this integral converges for an even m, one has the upper bounds

$$\mu_n^m \leqq \mu_n'^m + \int_D K^{(m)}(x, x) \, dx - \sum_{j=1}^{k} \mu_j'^m. \tag{12.3}$$

He also showed that one could obtain arbitrarily close upper bounds for μ_n by considering only the values of $\int K^{(m)}(x, x) \, dx$ for a sequence of values of m.

A simple method of this kind was given by Pólya (1968).

The principal difficulty with the Trefftz result was that it could only be applied when the kernel $K(x, y)$ which represents the operator T is explicitly known. Fichera (1965), (1966) has shown how to overcome this difficulty for many boundary value problems in partial differential equations. The idea is the following.

We consider the eigenvalue problem

$$\begin{aligned} Au &= \lambda u \quad \text{in } D, \\ u &\in V, \end{aligned} \tag{12.4}$$

where A is a symmetric elliptic operator of the form (1.6), Chap. 3, and the boundary conditions of V are such that if

$$(u, v) = \int u\bar{v}\, dx = \mathscr{B}(u, v),$$

then $\mathscr{A}(u, v) = (u, Av)$ is Hermitian and $\mathscr{A}(u, u)$ is positive definite.

If the boundary of D and the coefficients of A and of the operators which specify the boundary conditions are sufficiently regular and if the assumptions of Example 1.5 of Chap. 3 are valid, then for each f in $H^0(D)$ the problem

$$(12.5) \qquad \begin{gathered} Au = f \quad \text{in } D, \\ u \in V, \end{gathered}$$

has a generalized solution

$$u = Gf$$

which satisfies

$$(12.6) \qquad \begin{gathered} \mathscr{A}(u, v) = (f, v) \quad \text{for all } v \in V, \\ u \in V_{\mathscr{A}}. \end{gathered}$$

The solution operator G is Hermitian, and it can be represented as an integral operator

$$(12.7) \qquad Gu(x) = \int_D G(x, y) f(y)\, dy.$$

The Hermitian kernel $G(x, y)$ is called the Green's function of the problem (12.5). Unfortunately, it is usually not known. Suppose, however, that one does know a fundamental solution $F(x, y)$ of the operator A which satisfies the unstable boundary conditions. By this we mean that the transformation

$$(12.8) \qquad Ff(x) \equiv \int_D F(x, y) f(y)\, dy$$

has the properties

$$(12.9a) \qquad f \in H^0(D) \Rightarrow Ff \in H^M(D),$$

$$(12.9b) \qquad \mathscr{A}(v, Ff) = (v, f) \quad \text{for all } v \in V.$$

($2M$ is the order of A.)

Since $u = Gf$ is the solution of (12.5), G also satisfies

$$\mathscr{A}(v, Gf) = (v, f) \quad \text{for all } v \in V.$$

However, G has the additional property that Gf is in $V_{\mathscr{A}}$. Let $H_{\mathscr{A}}$ be the Hilbert space whose elements are the functions in $H^M(D)$ and whose scalar product is $\mathscr{A}(u, v)$. (Note that this scalar product is defined on all of $H^M(D)$.) Let P be the orthogonal projection in $H_{\mathscr{A}}$ onto the orthogonal complement of V. Then $Ff - PFf$ is in $V_{\mathscr{A}}$, and

$$\mathscr{A}(v, Ff - PFf) = \mathscr{A}(v, Ff) = (v, f) \quad \text{for all } v \in V.$$

Thus the Green's operator G has the representation

$$G = F - PF. \tag{12.10}$$

In order to obtain an integral operator on $H^0(D)$, we define the adjoint F^* of F as the transformation from $H_{\mathscr{A}}$ to $H^0(D)$ which makes the relation

$$(F^*u, f) = \mathscr{A}(u, Ff) \quad \text{for all } u \in H_{\mathscr{A}}, \quad f \in H^0(D)$$

valid. Because of (12.9b), we see that if v is in V, then F^*v is just the function v considered as a function of $H^0(D)$ instead of $H_{\mathscr{A}}$. In particular, this is true of G^*. Thus the eigenvalue problem (12.4) is equivalent to the problem

$$G^*Gu = \mu u, \tag{12.11}$$

where $\mu = 1/\lambda$. Because of (12.10) we easily find that

$$G^*G = F^*F - F^*PF. \tag{12.12}$$

This shows that the difference $F^*F - G^*G$ is positive semidefinite. Therefore the eigenvalues $\mu_1^{(0)} \geqq \mu_2^{(0)} \geqq \cdots$ of the operator

$$T_0 \equiv F^*F$$

are upper bounds for those of

$$T \equiv G^*G.$$

An interchange in the order of integration in the formula

$$(g, F^*Ff) = \mathscr{A}(Fg, Ff)$$

shows that T_0 can be written as an integral operator with the kernel

$$K_0(x, y) \equiv \mathscr{A}(F(\cdot, x), F(\cdot, y)).$$

An integration by parts shows that this kernel is $F(x, y)$ plus a boundary integral. In particular, the kernel for $T = G^*G$ is just $G(x, y)$.

For example, if we consider the eigenvalue problem

$$-\Delta u = \lambda u \quad \text{in } D,$$

$$u = 0 \quad \text{on } \partial D,$$

in three dimensions, we can take

$$F(x, y) = \frac{1}{4\pi|x - y|}.$$

Then

$$K_0(x, y) = \frac{1}{4\pi|x - y|} + \frac{1}{16\pi^2}\int_{\partial D} \frac{1}{|x - z|}\frac{\partial}{\partial n_z}\frac{1}{|y - z|}\,dS_z.$$

In general, one can show that for sufficiently large m the iterated kernel $K_0^{(m)}(x, y)$ has the property that

$$\int K_0^{(m)}(x, x)\,dx = \sum \mu_i^{(0)m}$$

converges. It gives an upper bound for $\int G^{(m)}(x, x)\,dx$. If we substitute this upper bound in (12.3), we obtain upper bounds for the eigenvalues μ_i of T.

These upper bounds are good only if $\int [K_0^{(m)}(x, x) - G^{(m)}(x, x)]\,dx$ is small. If this is not the case, we can improve the result by using the formula (12.12). We replace the uncomputable operator P by the operator P_l which is the projection onto the space spanned by l arbitrarily chosen functions $\phi_1, \cdots, \phi_l$ in the orthogonal complement of V in $H_{\mathscr{A}}$. If

$$\mathscr{A}(\phi_i, \phi_j) = \delta_{ij},$$

we have

$$P_l u = \sum_{i=1}^{l} \mathscr{A}(u, \phi_i)\phi_i.$$

Then the operator

$$T_l u \equiv F^*F - F^*P_lF$$

is an integral operator with kernel

$$K_l(x, y) = \mathscr{A}(F(\cdot, x), F(\cdot, y)) - \sum_{i=1}^{l} \mathscr{A}(F(\cdot, x), \phi_i)\mathscr{A}(\phi_i, F(\cdot, y)).$$

The eigenvalues $\mu_i^{(l)}$ of T_l satisfy $\mu_i \leqq \mu_i^{(l)} \leqq \mu_i^{(0)}$. Thus $\int K_l^{(m)}(x, x)\,dx$ gives a better upper bound for $\int G^{(m)}(x, x)\,dx$. Substituting $K_l^{(m)}$ for $K^{(m)}$ in (12.3), we obtain better upper bounds for the μ_n. If the ϕ_i are complete on the orthogonal complement of V, then $\int K_l^{(m)}(x, x)\,dx$ converges to $\int G^{(m)}(x, x)\,dx$ as $l \to \infty$.

If the operator A has constant coefficients, it is not too difficult to find a fundamental solution $F(x, y)$. If the coefficients are variable, Fichera (1971) has suggested the following alternative:

One finds an elliptic operator A_0 whose fundamental solution $F_0(x, y)$ is known explicitly and for which

$$0 < \mathscr{A}_0(u, u) \leqq \mathscr{A}(u, u) \quad \text{for all } u \in H^M(D), \quad u \neq 0.$$

The fundamental solution operator is to satisfy

$$\mathscr{A}_0(v, F_0 f) = (v, f) \quad \text{for all } v \in V.$$

Thus, it satisfies the unstable boundary conditions associated with the closure $V_{\mathscr{A}_0}$ and the bilinear form $\mathscr{A}_0(v, w)$.

Let $H_{\mathscr{A}_0}$ be the Hilbert space $H^M(D)$ with the new scalar product $\mathscr{A}_0(u, v)$. It is easily seen that if F_0^* is the adjoint of F_0 with respect to the scalar product $\mathscr{A}_0(u, v)$, and if P_0 is the projection orthogonal to V in this scalar product, then

$$(F_0^*F_0 - F_0^*P_0F_0) - G^*G$$

is positive definite. Therefore, the eigenvalues of $F_0^*F_0 - F_0^*P_0F_0$ lie above those of G^*G. Then clearly the eigenvalues of $F_0^*F_0^*$ lie above those of G^*G. Moreover, we can as before replace P_0 by an orthogonal projection P_l into the space spanned by the elements $\phi_1, \cdots, \phi_l$ of the orthogonal complement of V in $H_{\mathscr{A}_0}$. In this way we obtain a sequence

$$S_l = F_0^*F_0 - F_0^*P_lF_0 \tag{12.13}$$

which converges to $F_0^*F_0 - F_0^*P_0F_0$ but not to G^*G. To get convergence to G^*G, we define the positive semidefinite quadratic functional

$$\mathscr{A}'(u,u) = \mathscr{A}(u,u) - \mathscr{A}_0(u,u).$$

We define Q_r to be the orthogonal projection with respect to the scalar product $\mathscr{A}'(u,v)$ into the space spanned by the elements $\zeta_1, \zeta_2, \cdots, \zeta_r$ of $H_{\mathscr{A}_0}$. (We must assume that the ζ_i are so chosen that $\mathscr{A}'(\sum c_i\zeta_i, \sum c_i\zeta_i) = 0$ implies $\sum c_i\zeta_i = 0$.)

Then the quadratic functional

$$\mathscr{A}_r(u,u) \equiv \mathscr{A}_0(u,u) + \mathscr{A}'(Q_ru, Q_ru)$$

satisfies

$$\mathscr{A}_0(u,u) \leqq \mathscr{A}_r(u,u) \leqq \mathscr{A}(u,u).$$

Fichera showed that the Green's operator G_r which satisfies

$$\mathscr{A}_r(v, G_rf) = (v,f) \quad \text{for all} \quad v \in V,$$
$$G_rf \in V_{\mathscr{A}_r}$$

has the form

$$G_rf = G_0f - \sum_{i,j=1}^{r} \sigma_{ij}(f, \zeta_i - G_0A\zeta_i)[\zeta_j - G_0A\zeta_j],$$

where σ_{ij} is the inverse of the matrix $(\zeta_i - G_0A\zeta_i, A\zeta_j)$ and G_0 is the Green's operator for $\mathscr{A}_0$ on V.

For G_0 we again have the formula

$$G_0 = G_0^*G_0 = F_0^*F_0 - F_0^*P_0F_0,$$

where P_0 is the projection in $H_{\mathscr{A}_0}$ orthogonal to V. We now put

$$T_{rl} \equiv S_l - \sum \sigma_{ij}^{(l)}(f, \zeta_i - S_lA\zeta_i)[\zeta_j - S_lA\zeta_j],$$

where $\sigma_{ij}^{(l)}$ is the inverse of the matrix $(\zeta_i - S_lA\zeta_i, A\zeta_j)$ and S_l is defined by (12.13).

A combination of the first and second monotonicity principles shows that each eigenvalue $\mu_n^{(rl)}$ is nonincreasing in r and l and lies above μ_n. Moreover, T_{rl} can be expressed as an integral operator

$$T_{rl}f = \int K_{rl}(x,y)f(y)\,dy.$$

For sufficiently large m, $\int K_{rl}^{(m)}(x, x)\,dx$ is finite and provides an upper bound for $\int G^{(m)}(x, x)\,dx$. If we substitute it in (12.3), we obtain an upper bound for μ_n^m.

The constructions in the above methods are very much like those in the Weinstein and Aronszajn methods. However, the needed information and the actual computations involved are quite different.

One does not need to know the eigenvalues, eigenvectors, or resolvent of the base operator. One does need a fundamental solution.

One does not need to find the zeros of the Weinstein determinant. Instead one needs to compute a good approximation to an interated integral whose integrand is singular, and many good Rayleigh–Ritz lower bounds.

The integral $\int K^{(m)}(x, x)\,dx$ is just the trace of the operator T^m. Fichera has shown that similar approximation procedures can be based on the higher Fredholm invariants instead of the trace.

CHAPTER 5

Eigenvector Approximation

1. Error bounds for eigenvectors. So far we have only shown how to obtain approximations for the eigenvalues. It is frequently of as much or more interest to approximate the eigenvectors.

We observe that the error bound (2.3) of Chap. 4 shows that the error in the approximate eigenvalue behaves like the square of the error in the eigenvector. Therefore we cannot expect the eigenvectors to be approximated as accurately as the eigenvalues.

We begin with a simple bound of Trefftz (1933) for the error made in approximating u_1 by the Rayleigh–Ritz method. We choose some functions $v_1, \cdots, v_k$ in $V_{\mathscr{A}}$ and maximize the quotient $\mathscr{B}(\sum_{i=1}^k c_i v_i, \sum_{i=1}^k c_i v_i)/\mathscr{A}(\sum_{i=1}^k c_i v_i, \sum_{i=1}^k c_i v_i)$. We find a number

$$\mu_1' = \mathscr{B}(w_1, w_1)/\mathscr{A}(w_1, w_1),$$

which is a lower bound for the eigenvalue μ_1. We normalize w_1 so that

$$\mathscr{A}(w_1, w_1) = 1$$

and ask how close w_1 is to the correct normalized eigenfunction u_1.

Let us suppose that we also compute a lower bound μ_2' for μ_2 by the Rayleigh–Ritz method and that by some method or other we obtain upper bounds $\tilde{\mu}_1$ and $\tilde{\mu}_2$ for μ_1 and μ_2. We suppose that the bounds μ_1' and $\tilde{\mu}_2$ are so good that

$$\tilde{\mu}_2 < \mu_1'. \tag{1.1}$$

We write

$$w_1 = \mathscr{A}(w_1, u_1)u_1 + q,$$

where q is orthogonal to u_1. Then

$$\begin{aligned}\mu_1' - \tilde{\mu}_2 &= \mathscr{B}(w_1, w_1) - \tilde{\mu}_2\mathscr{A}(w_1, w_1) \\ &= (\mu_1 - \tilde{\mu}_2)|\mathscr{A}(w_1, u_1)|^2 + \mathscr{B}(q, q) - \tilde{\mu}_2\mathscr{A}(q, q).\end{aligned}$$

By the recursive definition of eigenvalues

$$\mathscr{B}(q, q) \leqq \mu_2\mathscr{A}(q, q) \leqq \tilde{\mu}_2\mathscr{A}(q, q).$$

Hence

$$\mu_1' - \tilde{\mu}_2 \leqq (\mu_1 - \tilde{\mu}_2)|\mathscr{A}(w_1, u_1)|^2 \leqq (\tilde{\mu}_1 - \tilde{\mu}_2)|\mathscr{A}(w_1, u_1)|^2$$

or

$$|\mathscr{A}(w_1, u_1)|^2 \geqq \frac{\mu_1' - \tilde{\mu}_2}{\tilde{\mu}_1 - \tilde{\mu}_2} = 1 - \frac{\tilde{\mu}_1 - \mu_1'}{\tilde{\mu}_1 - \tilde{\mu}_2}. \tag{1.2}$$

Because of our normalization,

$$\mathscr{A}(w_1 - u_1, w_1 - u_1) = 2 - 2\,\mathrm{Re}\,\mathscr{A}(w_1, u_1).$$

We assume that u_1 is normalized so that $\mathscr{A}(w_1, u_1)$ is real and positive. Then (1.2) can be written as

$$\mathscr{A}(w_1 - u_1, w_1 - u_1) \leqq 2\left[1 - \left\{1 - \frac{\tilde{\mu}_1 - \mu_1'}{\tilde{\mu}_1 - \tilde{\mu}_2}\right\}^{1/2}\right]. \tag{1.3}$$

This shows that if the ratio $(\tilde{\mu}_1 - \mu_1')/(\tilde{\mu}_1 - \tilde{\mu}_2)$ is small, then u_1 is near w_1 in norm.

If (1.1) is violated, the bound (1.2) becomes trivial. There is a very good reason for this. If μ_1 is a multiple eigenvalue, there is always an eigenvector u_1 which is orthogonal to w_1. That is, w_1 cannot possibly approximate all the eigenvectors which correspond to μ_1. If (1.1) is violated, there is no computational way of knowing whether or not μ_1 is a simple eigenvalue. In effect, it is multiple.

If μ_1 is multiple, we can ask whether a Rayleigh–Ritz eigenvector w_n lies close to some corresponding eigenvector. We shall do the same if μ_1 is only approximately multiple. Suppose, then, that for some $l \geqq 1$ and some $n \leqq l$ we have an upper bound $\tilde{\mu}_{l+1}$ for μ_{l+1} which is so good that $\tilde{\mu}_{l+1} < \mu_n'$ and that we have some upper bound $\tilde{\mu}_1$ for μ_1.

We write the Rayleigh–Ritz eigenfunction w_n which gives μ_n' as

$$w_n = \sum_{i=1}^{l} \mathscr{A}(w_n, u_i)u_i + q.$$

Then if $\mathscr{A}(w_n, w_n) = 1$,

$$\begin{aligned} \mu_n' - \tilde{\mu}_{l+1} &= \mathscr{B}(w_n, w_n) - \tilde{\mu}_{l+1}\mathscr{A}(w_n, w_n) \\ &= \sum_{j=1}^{l} (\mu_j - \tilde{\mu}_{l+1})|\mathscr{A}(w_n, u_j)|^2 + \mathscr{B}(q, q) - \tilde{\mu}_{l+1}\mathscr{A}(q, q) \\ &\leqq \sum_{j=1}^{l} (\mu_j - \tilde{\mu}_{l+1})|\mathscr{A}(w_n, u_j)|^2 \end{aligned} \tag{1.4}$$

since q is orthogonal to $u_1, \cdots, u_l$.

Since $\mu_j \leqq \tilde{\mu}_1$, we find that

$$\sum_{j=1}^{l} |\mathscr{A}(w_n, u_j)|^2 \geqq \frac{\mu_n' - \tilde{\mu}_{l+1}}{\tilde{\mu}_1 - \tilde{\mu}_{l+1}} = 1 - \frac{\tilde{\mu}_1 - \mu_n'}{\tilde{\mu}_1 - \tilde{\mu}_{l+1}}.$$

If we define the unit vector

$$u = \sum_{i=1}^{l} \mathscr{A}(w_n, u_i)u_i \Big/ \left\{\sum_{i=1}^{l} |\mathscr{A}(w_n, u_i)|^2\right\}^{1/2}$$

which is parallel to the projection of w_n onto the space spanned by $u_1, \cdots, u_n$, we can write this inequality as

$$\mathscr{A}(w_n - u, w_n - u) \leqq 2\left[1 - \left\{1 - \frac{\tilde{\mu}_1 - \mu'_n}{\tilde{\mu}_1 - \tilde{\mu}_{l+1}}\right\}^{1/2}\right]. \tag{1.5}$$

Thus if the ratio $(\tilde{\mu}_1 - \mu'_n)/(\tilde{\mu}_1 - \tilde{\mu}_{l+1})$ is small, then there is a linear combination u of the first l eigenvectors of norm one which is approximated by w_n.

We now extend this estimate to the case of a higher eigenvalue. We shall allow for the possibility that the eigenvalues $\mu_{m+1}, \mu_{m+2}, \cdots, \mu_l$ are equal or almost equal. We wish to show that if we have good Rayleigh–Ritz lower bounds $\mu'_1 \geqq \mu'_2 \geqq \cdots \geqq \mu'_m$ for $\mu_1 \geqq \cdots \geqq \mu_m$ and a good Rayleigh–Ritz lower bound μ'_n for u_n with $m + 1 \leqq n \leqq l$, and if there are known to be gaps between μ_m and μ_{m+1} and between μ_l and μ_{l+1}, then the Rayleigh–Ritz eigenvector w_n approximates a linear combination of the eigenvectors $u_{m+1}, \cdots, u_l$.

We first show that if $\mu'_1, \mu'_2, \cdots, \mu'_m$ are good approximations to the corresponding eigenvalues, then $u_1, \cdots, u_m$ are close to linear combinations of the Rayleigh–Ritz eigenvectors $w_1, \cdots, w_m$. For any w_i with $i \leqq m$ we apply the inequality (1.4) with n replaced by i and l by m. Then if $\mathscr{A}(w_i, w_j) = \delta_{ij}$,

$$\mu'_i - \tilde{\mu}_{m+1} \leqq \sum_{j=1}^{m} (\mu_j - \tilde{\mu}_{m+1})|\mathscr{A}(w_i, u_j)|^2.$$

We sum this inequality from $i = 1$ to m and observe that if we let

$$P'u \equiv \sum_{i=1}^{m} \mathscr{A}(u, w_i)w_i,$$

then because the w_i are orthonormal

$$\mathscr{A}(u - P'u, u - P'u) = \mathscr{A}(u, u) - \sum_{i=1}^{m} |\mathscr{A}(u, w_i)|^2.$$

We thus find

$$\sum_{i=1}^{m} (\mu'_i - \tilde{\mu}_{m+1}) \leqq \sum_{j=1}^{m} (\tilde{\mu}_j - \tilde{\mu}_{m+1})[1 - \mathscr{A}(u_j - P'u_j, u_j - P'u_j)]$$

or

$$\sum_{j=1}^{m} (\tilde{\mu}_j - \tilde{\mu}_{m+1})\mathscr{A}(u_j - P'u_j, u_j - P'u_j) \leqq \sum_{j=1}^{m} (\tilde{\mu}_j - \mu'_j). \tag{1.6}$$

We now consider w_n with $m + 1 \leqq n \leqq l$. Then w_n is orthogonal to $w_1, \cdots, w_m$. Therefore by Schwarz's inequality and (1.6),

$$\begin{aligned}\sum_{j=1}^{m} (\tilde{\mu}_j - \tilde{\mu}_{m+1})|\mathscr{A}(w_n, u_j)|^2 &= \sum_{j=1}^{m} (\tilde{\mu}_j - \tilde{\mu}_{m+1})|\mathscr{A}(w_n, u_j - P'u_j)|^2 \\ &\leqq \sum_{j=1}^{m} (\tilde{\mu}_j - \tilde{\mu}_{m+1})\mathscr{A}(u_j - P'u_j, u_j - P'u_j) \\ &\leqq \sum_{j=1}^{m} (\tilde{\mu}_j - \mu'_j).\end{aligned} \tag{1.7}$$

Substituting (1.7) in the inequality (1.4), we find that for $m + 1 \leqq n \leqq l$,

$$\mu'_n - \tilde{\mu}_{l+1} \leqq \max_{1 \leqq j \leqq m} \frac{\tilde{\mu}_j - \tilde{\mu}_{l+1}}{\tilde{\mu}_j - \tilde{\mu}_{m+1}} \sum_{i=1}^{m} (\tilde{\mu}_i - \mu'_i) + (\tilde{\mu}_{m+1} - \tilde{\mu}_{l+1}) \sum_{j=m+1}^{l} |\mathscr{A}(w_n, u_j)|^2$$

or

$$\sum_{j=m+1}^{l} |\mathscr{A}(w_n, u_j)|^2 \geqq \frac{1}{\tilde{\mu}_{m+1} - \tilde{\mu}_{l+1}} \left\{ \mu'_n - \tilde{\mu}_{l+1} - \frac{\tilde{\mu}_m - \tilde{\mu}_{l+1}}{\tilde{\mu}_m - \tilde{\mu}_{m+1}} \sum_{i=1}^{m} (\tilde{\mu}_i - \mu'_i) \right\}.$$

We now define the unit vector

$$u = \sum_{j=m+1}^{l} \mathscr{A}(w_n, u_i) u_i \Big/ \left\{ \sum_{i=1}^{m} |\mathscr{A}(w_n, u_i)|^2 \right\}^{1/2}$$

and show as before that

$$(1.8) \quad \begin{aligned} &\mathscr{A}(w_n - u, w_n - u) \\ &\leqq 2\left[1 - \left\{ 1 - \frac{1}{\tilde{\mu}_{m+1} - \tilde{\mu}_{l+1}} \left(\tilde{\mu}_{m+1} - \mu'_n + \frac{\tilde{\mu}_m - \tilde{\mu}_{l+1}}{\tilde{\mu}_m - \tilde{\mu}_{m+1}} \sum_{j=1}^{m} (\tilde{\mu}_j - \mu'_j) \right) \right\}^{1/2} \right]. \end{aligned}$$

This inequality shows that if μ'_n is close to $\tilde{\mu}_{m+1}$ relative to the spacing $\tilde{\mu}_{m+1} - \tilde{\mu}_{l+1}$ and if $\sum_{j=1}^{m} (\tilde{\mu}_j - \mu'_j)$ is small relative to $\tilde{\mu}_{m+1} - \tilde{\mu}_{l+1}$ and to $\tilde{\mu}_m - \tilde{\mu}_{m+1}$, then w_n is close to a linear combination of $u_{m+1}, \cdots, u_l$ of norm one. The simplest case is $m = l - 1$, when the eigenvalue is simple.

The author (1960) has obtained an inequality of the form (1.8) which is optimal in the following sense: The lower bound for $\mathscr{A}(w_n - u, w_n - u)$ depends upon finitely many upper bounds $\tilde{\mu}_i$ and lower bounds μ'_i, and for any such given values there is an operator whose eigenvalues lie between these upper and lower bounds and for which equality is attained in the error estimate. The optimal error estimate is rather cumbersome, and we state only a weakened version which is found in the same paper.

THEOREM 1.1. *If $m + 1 \leqq n \leqq l$, if $\mu'_m > \tilde{\mu}_{m+1}$, and if $\mu'_l > \tilde{\mu}_{l+1}$, then there is a linear combination u of norm one of the eigenvectors $u_{m+1}, \cdots, u_l$ which is approximated by the Rayleigh–Ritz eigenvector w_n in the sense that*

$$(1.9) \quad \begin{aligned} &\mathscr{A}(w_n - u, w_n - u) \\ &\leqq 2\left[1 - \left\{ 1 - \frac{\tilde{\mu}_{m+1} - \mu'_n}{\tilde{\mu}_{m+1} - \tilde{\mu}_{l+1}} \right\}^{1/2} \left\{ 1 - \frac{(\tilde{\mu}_{m+1} - \mu'_n)(\tilde{\mu}_1 - \mu'_m)}{(\mu'_m - \mu'_n)(\tilde{\mu}_1 - \tilde{\mu}_{m+1})} \right\}^{1/2} \right]. \end{aligned}$$

We shall not reproduce the proof of this result. We remark that the inequality (1.9) has the advantage over (1.8) that if $m > 2$, fewer upper and lower bounds need to be computed.

An inequality similar to (1.8) but without explicit constants was given by Birkhoff, de Boor, Swartz, and Wendroff (1966) and improved by Pierce and Varga (1972b).

Other eigenvector approximations have been given by Trefftz (1933), Bramble and Payne (1963), Moler and Payne (1968) and by Davis and Kahan (1970).

CHAPTER 6

Finite Difference Equations

1. Finite difference equations from the Rayleigh–Ritz method. The Rayleigh–Ritz method reduces the problem of approximating the eigenvalues $\mu_1 \geqq \mu_2 \geqq \cdots$ to that of computing the eigenvalues of $\mu'_1 \geqq \mu'_2 \geqq \cdots$ of the matrix problem

$$\sum_{j=1}^{k} \mathscr{B}(v_i, v_j)z_j = \mu \sum_{j=1}^{k} \mathscr{A}(v_i, v_j)z_j, \qquad i = 1, \cdots, k. \tag{1.1}$$

In principle this problem is equivalent to solving the polynomial equation

$$\det [\mathscr{B}(v_i, v_j) - \mu\mathscr{A}(v_i, v_j)] = 0.$$

However, when k is not sufficiently small, the numerical computation of a determinant is not only time-consuming, but it also involves so many calculations that the roundoff error swamps the desired result unless inordinate precision is used. Moreover, other ways of approximating the eigenvalues of (1.1) are, in general, also unsuccessful when k is too large. We thus arrive at a dilemma: the error bounds of § 2 of Chap. 4 tell us to make k large in order to make the error $\mu_n - \mu'_n$ small, but for such a large k we cannot usually compute the μ'_n.

An alternative approach to the eigenvalue problem $Bu = \mu Au$ when A and B are ordinary or partial differential operators is to replace the differential problem by a finite difference problem. The idea is to consider mesh functions w_h which are defined only on a finite set of k points (the mesh points) and to replace the derivatives occurring in the operators A and B by some corresponding difference quotients. Then at each mesh point P the differential equation $Bu(P) = \mu Au(P)$ is replaced by an equation of the form

$$\sum b_\alpha(P)w_h(Q_\alpha) = \mu \sum a_\alpha(P)w_h(Q_\alpha). \tag{1.2}$$

The sums are over those mesh points Q_α which lie near P. The coefficients $a_\alpha(P)$ and $b_\alpha(P)$ are so chosen that for any infinitely differentiable function $v(x)$ in V

$$\sum a_\alpha v(Q_\alpha) \cong Av(P),$$
$$\sum b_\alpha v(Q_\alpha) \cong Bv(P).$$

This approximation is good provided the maximum distance from any Q_α in the sum to P is sufficiently small. The boundary data are to be taken into account for mesh points near or on the boundary.

The system (1.2) for all mesh points P is again in the form of a matrix eigenvalue problem

$$B_h \mathbf{w}_h = \mu A_h \mathbf{w}_h. \tag{1.3}$$

If we are not careful in setting up the finite difference equation, the coefficient matrices may be nonsymmetric, so that the eigenvalues μ may not even be real. What, then, is the advantage of the linear system (1.3) over the linear system (1.1)?

The answer is a purely computational one. Each sum in (1.2) extends over only a small number (relative to k) of points Q_α which are near P. This means that the matrices B_h and A_h have the property that all but a few of the elements of each row are zero. Such matrices are said to be *sparse*.

Sparse $k \times k$ matrices have several computational advantages. In the first place, if each row has at most l nonzero elements, it is only necessary to store roughly $2kl$ pieces of data (the l nonzero entries of each row and their locations) rather than the k^2 pieces of data needed for a full $k \times k$ matrix. Second, the number of computations needed in an algorithm to determine eigenvalues also is proportional to kl rather than to k^2. It follows that for a given amount of computer storage and computer time the size k of the matrices for which the eigenvalues can be determined is much larger if the matrices are sparse than if they are not.

It was pointed out by Courant (1927), (1943) that the Rayleigh–Ritz method (1.1) will produce sparse matrices if the trial functions v_i are chosen judiciously. Courant (1943) demonstrated one such choice for a two-dimensional second order problem. He divided the domain into nonoverlapping triangles whose vertices are the mesh points. He then let the trial functions for the Rayleigh–Ritz method consist of piecewise linear functions which are continuous and whose derivatives have discontinuities only across the edges of the triangles. Such a function is determined by its values at the mesh points. If we let v_i be the function which is one at the ith mesh point and zero at all the others, its graph is a pyramid. (Synge (1957) called such a function a pyramid function.) More important, v_i is identically zero except on those triangles of which the ith mesh point is a vertex.

Therefore, $\mathscr{A}(v_i, v_j) = \mathscr{B}(v_i, v_j) = 0$ unless the ith and jth mesh points are vertices of the same triangle. That is, the entries in the ith row of $\mathscr{A}(v_i, v_j)$ and $\mathscr{B}(v_i, v_j)$ are zero except for those j for which there is a line connecting the ith and jth mesh points. If we connect each mesh point to at most l others, then no matter how many mesh points there are, the matrices in (1.1) are sparse.

Example 1.1 (Pólya (1952)).

$$-\Delta u = \lambda u \quad \text{in } D,$$
$$u = 0 \quad \text{on } \partial D, \tag{1.4}$$

where D is a bounded two-dimensional domain.

Introduce the mesh points $x = rh$, $y = sh$, $r, s = 0, \pm 1, \pm 2, \cdots$, where h is a small parameter (the mesh size). Connect the vertex (rh, sh) to the vertices $((r \pm 1)h, sh)$, $(rh, (s \pm 1)h)$, $((r + 1)h, (s + 1)h)$, and $((r - 1)h, (s - 1)h)$. For (r, s) such that all the triangles having (rh, sh) as a vertex lie in $D \cup \partial D$, define $v_{rs}(x)$ to be the piecewise linear function which is one at (rh, sh) and 0 at the other mesh points.

A computation shows that (1.1) becomes

$$
\begin{aligned}
&\tfrac{1}{12}\{6w(rh,sh) + w((r+1)h,sh) + w((r-1)h,sh) + w(rh,(s+1)h) \\
&\quad + w(rh,(s-1)h) + w((r+1)h,(s+1)h) + w(r-1)h,(s-1)h)\} \\
&\quad = \mu h^{-2}\{4w(rh,sh) - w((r+1)h,sh) - w((r-1)h,sh) \\
&\qquad - w(rh,(s+1)h) - w(rh,(s-1)h)\}
\end{aligned} \tag{1.5}
$$

with the convention that $w(rh, sh) = 0$ if (rh, sh) is a vertex of a triangle that goes outside $D \cup \partial D$.

The right-hand side is a standard finite differences approximation to $-\mu\,\Delta w$. The left-hand side is not standard, but it clearly approximates $w(rh, sh)$.

The basic property of the v_i that makes the matrices sparse is the fact that each v_i vanishes outside some set S_i and only a relatively small number of S_i intersect. Whenever we have a set of trial functions v_i with the property that each support S_i intersects at most l of the S_j with l small, the matrices in (1.1) are sparse. Thus we obtain a system (1.1) which has the important property of finite difference equations and which gives lower bounds $\mu_n^{(h)}$ for the eigenvalues.

This idea, then, allows one to obtain a finite difference scheme whose eigenvalues $\mu_n^{(h)}$ are lower bounds for the eigenvalues μ_n of the original problem. It is not clear, however, that it is always computationally desirable to proceed in this fashion.

A human being with good intuition can frequently guess the general shape of the first few eigenfunctions u_i. If one uses these guesses as trial functions v_i in the Rayleigh–Ritz method, one may well obtain better lower bounds for the first few eigenvalues with a small number k of v_i than one gets from the difference problem with a large number k_h of mesh points. This is because the trial functions used in deriving the finite difference scheme are usually far from the eigenfunctions. In a case like this there may be less computation involved in solving the problem (1.1) with full $k \times k$ matrices than in solving it with sparse $k_h \times k_h$ matrices.

In general, one would expect to be able to combine the above ideas by allowing the support of a few trial functions to be the whole domain while the remainder have small support. The result will be matrices in (1.1) which have only l nonzero elements per row except for a small number l' of rows.

The above ideas provide lower bounds for the μ_i. It is still necessary to obtain upper bounds either through the error bounds of § 2 of Chap. 4 or by some other means.

2. The method of finite elements. In the preceding section we showed how to produce a finite difference problem whose eigenvalues $\mu_n^{(h)}$ give lower bounds for the eigenvalues μ_n of the given differential problem. Since the method used is the Rayleigh–Ritz method, an error bound can be obtained from the results of § 2 of Chap. 4.

In order to facilitate high-speed computation of eigenvalues with a minimum of manual labor, it is useful to utilize a class of trial functions v_i which is more or less independent of the particular problem and to obtain error bounds which depend on the operators and the domain of the particular problem only through a multiplicative constant.

The search for such universal classes of trial functions and the corresponding error bounds has been named the method of finite elements.

The train of ideas is the following:

(i) Let A be an elliptic operator of order $2M$ of the form (1.6) of Chap. 3. By integration by parts as in Example 1.5 of Chap. 3 we see that there is a constant c_1 so that

$$\mathscr{A}(u,u) \leqq c_1 \|u\|^2_{H^M(D)} \equiv c_1 \sum_{\alpha_1+\cdots+\alpha_N \leqq M} \int_D \left| \frac{\partial^{\alpha_1+\cdots+\alpha_N} u}{\partial x_1^{\alpha_1}\cdots \partial x_N^{\alpha_N}} \right|^2 dx. \tag{2.1}$$

(ii) Suppose that the problem $Au = f$ is coercive, meaning that there is a positive constant c_2 so that

$$\|u\|^2_{H^M(D)} \leqq c_2 \mathscr{A}(u,u). \tag{2.2}$$

If the boundary of D, the coefficients which define A, and the boundary conditions are sufficiently smooth, it follows from the elliptic regularity theory (see, e.g., Fichera (1965)) that for any nonnegative integer J there is a constant c_3 which depends upon V, the coefficients of A, the boundary of D, and J so that $u \in V$ and $Av \in H^J(D)$ imply

$$\|u\|^2_{H^{J+2M}(D)} \leqq c_2 c_3 \|Au\|^2_{H^J(D)}. \tag{2.3}$$

An eigenfunction u_i corresponding to an eigenvalue $\mu_i \neq 0$ satisfies

$$Au_i = \frac{1}{\mu_i} Bu_i. \tag{2.4}$$

If B is of order $K < 2M$, then there is a constant c_4 depending only on the coefficients of B and on J such that

$$\|Bu_i\|^2_{H^J(D)} \leqq c_4 \|u_i\|^2_{H^{J+K}(D)}. \tag{2.5}$$

Putting (2.3), (2.4), and (2.5) together, we see that

$$\|u_i\|^2_{H^{J+2M}(D)} \leqq \frac{c_2 c_3 c_4}{\mu_i^2} \|u_i\|^2_{H^{J+K}(D)}. \tag{2.6}$$

In particular, if we let $J = M - K$, we see from (2.2) that

$$\|u_i\|^2_{H^{3M-K}(D)} \leqq \frac{c_2^2 c_3 c_4}{\mu_i^2} \mathscr{A}(u_i, u_i). \tag{2.7}$$

If we wish, we can now let $J = 3M - 2K$ in (2.6) to get a bound for an even higher Sobolev norm. Proceeding in this way, we find that

$$\|u_i\|_{H^{M+r(2M-K)}(D)} \leqq c_2 \left(\frac{c_2 c_3 c_4}{\mu_i^2} \right)^r \mathscr{A}(u_i, u_i). \tag{2.8}$$

In this way we can obtain bounds for an arbitrarily high Sobolev norm. However, c_3 and c_4 depend on r.

(iii) The inequalities (2.1) and (2.8) now imply that the estimate (2.1) of Chap. 4 will follow from a bound for the quantity

$$\kappa_1 \equiv \sup_{u\in V} \inf_{v=c_1v_1+\cdots+c_kv_k} \frac{\|u - v\|^2_{H^M(D)}}{\|u\|^2_{H^L(D)}},$$

with $L > M$.

We note that the infimum is attained when $v = P_k u$, the orthogonal projection in the scalar product of $H^M(D)$ onto the span of the v_i. Thus, we are left with the problem of estimating

$$\kappa_1 \equiv \sup_{u\in V} \frac{\|u - P_k u\|^2_{H^M(D)}}{\|u\|^2_{H^L(D)}}. \tag{2.9}$$

·This is another eigenvalue problem. We know from the separation theorem that κ_1 is bounded below by the $(k + 1)$st eigenvalue of the Rayleigh quotient $\|u\|^2_{H^M(D)}/\|u\|^2_{H^L(D)}$. We may be able to apply the methods of § 3 of Chap. 4 to find the eigenvalue κ_1.

The problem is thus reduced to finding a class of functions $v_1, \cdots, v_k$ of small support such that κ_1 is small. The classes of functions that are usually considered consist of piecewise polynomials with breaks in some derivatives along certain mesh planes.

In one dimension there are two principal classes of such functions: the splines and the Hermite interpolation functions.

The domain in one dimension is just an interval, say (0, 1), and the mesh points $0 = x_0 < x_1 < \cdots < x_k < x_{k+1} = 1$ divide this interval into $k + 1$ subintervals. Both classes of functions are indexed by a positive integer q. They are polynomials of degree $2q - 1$ in each subinterval.

The *Hermite interpolation functions* are required to have continuous derivatives up to order $q - 1$ at the mesh points, and boundary conditions may be prescribed at the ends. Corresponding to each mesh point x_j there are q linearly independent such functions $v_{ij}, i = 1, \cdots, q$, with the property that $v_{ij}(x) = 0$ for $x < x_{j-1}$ and $x > x_{j+1}$ and that

$$\frac{d^{2q-m}v_{ij}(x_j)}{dx^{2q-m}} = \delta_{im}, \qquad i, m = 1, \cdots, q.$$

Thus the maximal number l of v_{ij} for which $\mathscr{A}(v_{rs}, v_{ij}) \neq 0$ or $\mathscr{B}(v_{rs}, v_{ij}) \neq 0$ is $3q$. The index q must be taken to be at least M. The linear combinations of the v_{ij} together with the at most $2q$ additional polynomials which are nonzero only in the end intervals give the whole set of Hermite functions of index q.

The *spline functions* are required to have $2q - 2$ continuous derivatives at each mesh point, and again may be required to satisfy boundary conditions. For $q \leqq j \leqq k + 1 - q$ a unique spline function v_j can be constructed with the properties that $v_j \equiv 0$ for $x \leqq x_{j-q}$ and $x \geqq x_{j+q}$ and that $v_j(x_j) = 1$. The number l of v_j for which $\mathscr{A}(v_i, v_j) \neq 0$ or $\mathscr{B}(v_i, v_j) \neq 0$ is $4q - 1$. The set of splines of index q is spanned by these v_j plus some others which correspond to the points near the ends. We need to make $2q > M$.

The approximability results needed for the error bounds in § 2, Chap. 4, have been proved for these cases by Birkhoff and de Boor (1965) for $q = 2$, and for the general case by Ciarlet, Schultz, and Varga (1967). A very general class of approximation functions in one dimension and a good bibliography are given by Schultz and Varga (1967).

These results have been applied to obtain error bounds for the eigenvalues and eigenvectors of one-dimensional problems by Birkhoff, de Boor, Swartz and Wendroff (1966), Ciarlet, Schultz, and Varga (1968) and Strang and Fix (1973).

In higher dimensions there are added difficulties. The boundary of the domain may be complicated, and therefore functions which satisfy the stable boundary conditions are difficult to construct.

The problem is easiest in cases where there are no stable boundary conditions. One needs to assume that the domain has a boundary which is so smooth that every function of the Sobolev space $H^s(D)$ for some suitable s can be extended into a function of $H^s(R^N)$ on the whole space. That is, there exists a linear transformation $\pi : H^s(D) \to H^s(R^N)$, and a constant c such that the restriction of πu to D is u and

$$\|\pi u\|_{H^s(R^N)} \leqq c\|u\|_{H^s(D)}. \tag{2.10}$$

If this condition is satisfied, the problem of approximating functions in D can be reduced to that of approximating functions on R^N. For if πu is approximated by a linear combination of v_i on R^N, then u is approximated by the restriction to D of this linear combination.

These ideas have been worked out for approximations by products of piecewise polynomials by Aubin (1967), (1968) and by Schultz (1969c, b, d). Bramble and Hilbert (1970), (1971) have obtained approximability estimates on R^N. Simpler bounds were obtained when D is a polygon by Birkhoff (1969), Birkhoff and de Boor (1965), Birkhoff, Schultz, and Varga (1968), Bramble and Zlamal (1970), Carlson and Hall (1971), Hall (1969) and Zlamal (1968).

The only case of stable boundary conditions that has been treated is the extreme case where u and all partial derivatives up to order $M - 1$ vanish on the boundary. Schultz (1969a, c) has shown that one can approximate such a u in the norm $H^M(D)$ by a product of a piecewise polynomial of some index $q \geqq M$ with a function $\theta(x)^M$ where θ is a smooth function with $\theta > 0$ in D and $\theta = 0$ on the boundary. Aubin and Burchard (1971) have shown that one can also approximate such a u by spline functions which satisfy the boundary conditions.

Fix and Strang (1969), Strang and Fix (1973), and Babuska (1971) have considered a more general class of trial functions v_i which are formed from a small collection of functions $\phi_1, \cdots, \phi_r$ by applying translations and shrinking by a dilation factor $1/h$.

Error bounds for eigenvalues in multidimensional problems have been given by Schultz (1969a), Pierce and Varga (1972a), Birkhoff and Fix (1970), and Strang and Fix (1973). The books of Varga (1971) and Strang and Fix (1973) contain good summaries of these results.

All the above papers give error bounds of the form ch^t where h is the mesh size. The integer t can usually be found explicitly. The constant c depends upon the constants in (2.1), (2.2), (2.3) and (2.5), as well as the constant introduced when the functions are extended beyond the domain. In one dimension it is possible to give reasonable estimates for these constants and hence for c. In more dimensions it is almost impossible to find reasonable bounds for c, and the constant c may be huge. This fact is easily overlooked when one only states that the error is $O(h^t)$.

This notation gives the impression that the larger t is, the better the approximation method. However, the reader should keep two facts in mind. In the first place, the constant c in the error bound ch^t depends in a rather complicated fashion on t. If we replace $c(t)h^t$ by $c(t')h^{t'}$ with $t' > t$, it is true that for sufficiently small h the second bound is better. In fact, this occurs if

$$h < [c(t)/c(t')]^{1/(t'-t)}.$$

If $c(t)$ is large but $c(t')$ is even larger, a computation with such a small mesh size may well be beyond the capability of the available computational facilities.

There is a second adverse effect involved in increasing t. An increase in t must usually be accompanied by an increase in the number l of nonzero elements in the rows of the matrices $\mathscr{A}(v_i, v_j)$ and $\mathscr{B}(v_i, v_j)$. This adds considerably to the computational difficulty.

In order to make a fair comparison between two finite difference schemes we should compare the accuracies that can be obtained from a given amount of computation. We have seen that if t is increased and h is fixed the computational labor is larger for the second scheme. Hence to conserve computational effort we must increase the mesh size h. Thus the error bounds for two schemes are of the form $c(t)h^t$ and $c(t')h'^{t'}$ where when $t' > t$, $c(t') > c(t)$ and $h' > h$. It is unclear, then, whether a scheme with accuracy of order t' is in fact better than one with accuracy of order t.

In particular, it is preferable to have a scheme for which the constant $c(t)$ is known explicitly. In the next section we shall show how this can often be done.

3. Upper and lower bounds from finite difference equations. As the Example 1.1 showed, the finite difference approximation obtained by means of the Rayleigh–Ritz method may not be the simplest or most natural one. Moreover, if the operators A and B have variable coefficients, the computation of the matrices $\mathscr{A}(v_i, v_j)$ and $\mathscr{B}(v_i, v_j)$ may be a formidable task in itself, and it may not be possible to compute these matrices exactly.

It is therefore useful to estimate the error one makes in using an arbitrary finite difference approximation.

It was shown by Krylov (1928), (1931) that one can often find both upper and lower bounds for the eigenvalues $\mu_1 \geqq \mu_2 \geqq \cdots$ of a Rayleigh quotient $\mathscr{B}(u, u)/\mathscr{A}(u, u)$ on the space V in terms of the eigenvalues $\mu_1^{(k)} \geqq \mu_2^{(k)} \geqq \cdots$ of an approximating algebraic problem

$$\sum_{j=1}^{k} B_{ij}\bar{z}_j = \mu \sum_{j=1}^{k} A_{ij}\bar{z}_j, \qquad i = 1, \cdots, k, \tag{3.1}$$

provided the $k \times k$ matrices A_{ij} and B_{ij} are Hermitian and A_{ij} is positive definite.

We define the two quadratic functionals

$$\mathscr{A}_h(\mathbf{z}, \mathbf{z}) = \sum_{i,j=1}^{k} A_{ij} z_i \bar{z}_j,$$

$$\mathscr{B}_h(\mathbf{z}, \mathbf{z}) = \sum_{i,j=1}^{k} B_{ij} z_i \bar{z}_j$$

on the space Z_k of k-tuples $\{z_1, \cdots, z_k\}$. Let $\mu_1^{(h)} \geqq \mu_2^{(h)} \geqq \cdots$ be the eigenvalues of the Rayleigh quotient $\mathscr{B}_h(\mathbf{z}, \mathbf{z})/\mathscr{A}_h(\mathbf{z}, \mathbf{z})$. These eigenvalues are also the eigenvalues of the system (3.1).

From the point of view of computation it is useful to have the matrices A_{ij} and B_{ij} be sparse.

In order to obtain a lower bound for μ_n, one applies the mapping principle of § 6 of Chap. 3, with $\mathscr{A}_1 = \mathscr{A}_h$, $\mathscr{B}_1 = \mathscr{B}_h$, $\mathscr{A}_2 = \mathscr{A}$, $\mathscr{B}_2 = \mathscr{B}$ and with M a linear mapping M_h from Z_k to V. Such a mapping must be of the form

$$M_h(z_1, \cdots, z_k) = \sum_{i=1}^{k} z_i v_i,$$

where $v_1, \cdots, v_k$ are elements of V. The inequalities (6.1) and (6.2) of Chap. 3 become

$$\frac{\mathscr{B}(\sum z_i v_i, \sum z_i v_i)}{\mathscr{A}(\sum z_i v_i, \sum z_i v_i)} \geqq f\left(\frac{\sum A_{ij} z_i \bar{z}_j}{\sum B_{ij} z_i \bar{z}_j}\right), \tag{3.2}$$

$$\mathscr{A}(\sum z_i v_i, \sum z_i v_i) \geqq g\left(\frac{\sum A_{ij} z_i \bar{z}_j}{\sum B_{ij} z_i \bar{z}_j}\right) \sum A_{ij} z_i \bar{z}_j. \tag{3.3}$$

If one makes

$$A_{ij} = \mathscr{A}(v_i, v_j), \qquad B_{ij} = \mathscr{B}(v_i, v_j),$$

these inequalities are certainly satisfied with $f(\xi) = \xi$, $g(\xi) \equiv 1$, and we find the Rayleigh–Ritz inequality $\mu_n^{(h)} \leqq \mu_n$. This is the method discussed in § 1. However, the present method is applicable to other finite difference schemes: that is, to other matrices A_{ij} and B_{ij}.

Example 3.1 (Weinberger (1956)). We again consider the problem (1.4). A finite difference scheme that is much simpler than (1.5) is the standard scheme

$$\begin{aligned} w(rh, sh) = \mu h^{-2}\{&4w(rh, sh) - w((r+1)h, sh) - w((r-1)h, sh) \\ &- w(rh, (s+1)h) - w(rh, (s-1)h)\}, \end{aligned} \tag{3.4}$$

again with the convention that $w(rh, sh) = 0$ if (rh, sh) is a vertex of a triangle that extends outside $D \cup \partial D$. (We use here the more suggestive notation $w(rh, sh)$ for a mesh function instead of the numbers z_i.) It is easily seen by multiplying (3.4) by $v(rh, sh)$ and summing that the coefficient matrices involved are symmetric and positive definite. In fact, we have

$$\mathscr{B}_h(w, w) \equiv \sum w(rh, sh)^2 h^2, \tag{3.5}$$

$$\mathscr{A}_h(w, w) \equiv \sum \left\{ \left[\frac{w((r+1)h, sh) - w(rh, sh)}{h} \right]^2 + \left[\frac{w(rh, (s+1)h) - w(rh, sh)}{h} \right]^2 \right\} h^2.$$

We now let $M_h w$ be the continuous piecewise linear function whose value at the mesh point (rh, sh) is $w(rh, sh)$ and whose derivatives have discontinuities only on lines $x = ih$, $y = jh$ and $x - y = kh$ as in the Courant–Pólya scheme in § 1 of this chapter. We can write

$$M_h w = \sum w(rh, sh) v_{rs}(x, y)$$

where the v_{rs} are pyramid functions. Our boundary condition for w is equivalent to the condition $M_h w \in V$.

A computation shows that

$$\begin{aligned} \mathscr{B}(M_h w, M_h w) = \tfrac{1}{12} h^2 \sum \{ 12 w(rh, sh)^2 &- [w((r+1)h, sh) - w(rh, sh)]^2 \\ &- [w(rh, (s+1)h) - w(rh, sh)]^2 \\ &- [w(r+1)h, (s+1)h) - w(rh, sh)]^2 \}, \end{aligned} \tag{3.6}$$

while

$$\mathscr{A}(M_h w, M_h w) = \mathscr{A}_h(w, w).$$

(The Rayleigh quotient $\mathscr{B}(M_h w, M_h w)/\mathscr{A}(M_h w, M_h w)$ gives rise to the finite difference equation (1.5).)

We now observe that

$$\begin{aligned} [w((r+1)h, (s+1)h) - w(rh, sh)]^2 \leqq 2[w((r+1)h, (s+1)h) &- w(rh, (s+1)h)]^2 \\ &+ 2[w(rh, (s+1)h) - w(rh, sh)]^2. \end{aligned}$$

Hence we see from (3.6) that

$$\begin{aligned} \mathscr{B}(M_h w, M_h w) &\geqq \mathscr{B}_h(w, w) - \tfrac{1}{4} h^2 \mathscr{A}_h(w, w) \\ &= \left\{ \frac{\mathscr{B}_h(w, w)}{\mathscr{A}_h(w, w)} - \tfrac{1}{4} h^2 \right\} \mathscr{A}(M_h w, M_h w). \end{aligned}$$

We thus have the inequalities (6.1) and (6.2) of Chap. 3 with $f(\xi) = \xi - 3h^2$, $g(\xi) \equiv 1$. The mapping principle shows that if the finite difference scheme (3.4) is used, we obtain the lower bounds

$$\mu_n \geqq \mu_n^{(h)} - \tfrac{1}{4} h^2 \tag{3.7}$$

for the eigenvalues. For the eigenvalues $\lambda_n = 1/\mu_n$ of the problem (1.4) we then obtain the upper bounds

$$\lambda_n \leqq \frac{\lambda_n^{(h)}}{1 - \frac{1}{4}h^2\lambda_n^{(h)}} \tag{3.8}$$

provided h^2 is so small that $\frac{1}{4}h^2\lambda_n^{(h)} < 1$.

Thus the simpler scheme (3.4) is just as capable of giving lower bounds as the more complicated scheme (1.5), which is dictated by the Rayleigh–Ritz method.

Bounding procedures of the above kind for ordinary differential equations were already introduced by Krylov (1928), (1931).

In order to find an upper bound for the eigenvalue μ_n, we need to apply the mapping principle in the opposite direction. That is we must produce a mapping M from functions in V to the k-tuples of Z_k for which the estimates (6.1) and (6.2) of Chap. 3 can be established. Note that M can be considered as a k-tuple of linear functionals on V.

A simple mapping M is defined by letting Mv be the set of values of v at some given points (the mesh points) of the domain. In this case it is clear that $\mathscr{A}_h(Mv, Mv)$ and $\mathscr{B}_h(Mv, Mv)$ should be quadrature formulas which approximate $\mathscr{A}(v, v)$ and $\mathscr{B}(v, v)$, respectively.

This simple mapping works quite well in the case of ordinary differential equations, and has been applied by Krylov (1928), (1931) and by Wendroff (1965).

In the case of second order partial differential equations the value of the function v at a point is not a bounded linear functional in terms of $\mathscr{A}(v, v)$. Therefore it becomes preferable to let Mv be defined by averages of v in neighborhoods of the mesh points.

Example 3.2 (Weinberger (1958)). We again consider the problem of approximating the eigenvalues $\mu_i = 1/\lambda_i$ of (1.4) by those of the finite difference scheme (3.4).

We recall that the functions of V vanish on the boundary of D. We continue these functions as zero outside V, and define the mapping

$$(Mv)(rh, sh) \equiv h^{-2}\int_{-h/2}^{h/2}\int_{-h/2}^{h/2} v(rh + \xi, sh + \mu)\, d\xi\, d\mu. \tag{3.9}$$

We can be sure that this integral is zero if the square $S_{rs} \equiv \{x, y \mid |x - rh| \leqq \frac{1}{2}h, |y - sh| \leqq \frac{1}{2}h\}$ lies outside of D. We therefore make the convention that (3.4) holds if S_{rs} intersects D and that $w(rh, sh) = 0$ if S_{rs} does not intersect D. Note that we are now working on a larger mesh domain (that is, one with more interior points) than when we were seeking an upper bound for μ_n. We call the eigenvalues of the problem (3.4) on this large mesh domain $\tilde{\mu}_1^{(h)} \geqq \tilde{\mu}_2^{(h)} \geqq \cdots$.

An integration by parts shows that

$$Mu((r + 1)h, sh) - Mu(rh, sh) = h^{-2}\int_{-h/2}^{h/2}\int_{-h/2}^{3h/2} \frac{\partial u}{\partial x}(rh + \xi, sh + \eta)\psi(\xi)\, d\xi\, d\eta,$$

where

$$\psi(\xi) = \begin{cases} \xi + \frac{1}{2}h & \text{for } -\frac{1}{2}h \leqq \xi \leqq \frac{1}{2}h, \\ \frac{3}{2}h - \xi & \text{for } \frac{1}{2}h \leqq \xi \leqq \frac{3}{2}h, \\ 0 & \text{elsewhere}. \end{cases}$$

Therefore by Schwarz's inequality,

$$[Mu(r+1)h, sh) - Mu(rh, sh)]^2 \leqq h^{-4} \iint \psi(\xi)\, d\xi\, d\eta \iint \left|\frac{\partial u}{\partial x}\right|^2 \psi(\xi)\, d\xi\, d\eta$$

$$= h^{-1} \int_{-h/2}^{h/2} \int_{-h/2}^{3h/2} \left|\frac{\partial u}{\partial x}\right|^2 \psi(\xi)\, d\xi\, d\eta .$$

We add these inequalities and observe that $\psi(x-h) + \psi(x) + \psi(x+h) \equiv h$ to see that

$$\sum [Mu((r+1)h, sh) - Mu(rh, sh)]^2 \leqq \int_D\int \left|\frac{\partial u}{\partial x}\right|^2 dx\, dy.$$

The same computation for the y-differences shows that

$$\mathscr{A}_h(Mu, Mu) \leqq \mathscr{A}(u, u). \tag{3.10}$$

Because of the definition (3.9) we have

$$\begin{aligned} &\int_{-h/2}^{h/2}\int_{-h/2}^{h/2} [u(rh+\xi, sh+\eta) - Mu(rh, sh)]^2\, d\xi\, d\eta \\ &\quad = \int_{-h/2}^{h/2}\int_{-h/2}^{h/2} u(rh+\xi, sh+\eta)^2\, d\xi\, d\eta - h^2[Mu(rh, sh)]^2. \end{aligned} \tag{3.11}$$

Also because of (3.9)

$$\int_{-h/2}^{h/2}\int_{-h/2}^{h/2} [u(rh+\xi, sh+\eta) - Mu(rh, sh)]\, d\xi\, d\eta = 0. \tag{3.12}$$

The eigenvalues $\nu_1 \geqq \nu_2 \geqq \cdots$ of the Rayleigh quotient

$$\frac{\int_{-h/2}^{h/2}\int_{-h/2}^{h/2} v^2\, d\xi\, d\eta}{\int_{-h/2}^{h/2}\int_{-h/2}^{h/2} [|\text{grad } v|^2 + v^2]\, d\xi\, d\eta}$$

are those of the problem

$$v = \nu[-\Delta v + v] \quad \text{in } D_h\colon\ |\xi| \leqq \tfrac{1}{2}h,\ \ |\eta| \leqq \tfrac{1}{2}h,$$

$$\frac{\partial v}{\partial n} = 0 \qquad \text{on } \partial D_h.$$

This problem can be solved by separation of variables. We find that $\nu_1 = 1$ with the eigenfunction $v_1 \equiv 1$. The second eigenvalue

$$\nu_2 = \nu_3 = \frac{1}{\pi^2 h^{-2} + 1}$$

with the eigenfunctions $\sin \pi h^{-1} x$ and $\sin \pi h^{-1} y$. The condition (3.12) shows that the function

$$v \equiv u(rh + \xi, sh + \eta) - Mu(rh, sh) \tag{3.13}$$

is orthogonal to v_1. Therefore,

$$\int_{D_h}\int v^2 \, d\xi \, d\eta \leqq \nu_2 \int_{D_h}\int [|\text{grad } v|^2 + v^2] \, d\xi \, d\eta$$

or

$$\pi^2 h^{-2} \int_{D_h}\int v^2 \, d\xi \, d\eta \leqq \int_{D_h}\int |\text{grad } v|^2 \, d\xi \, d\eta.$$

Recalling the definition (3.13) and the identity (3.11), we obtain the inequality

$$\int_{-h/2}^{h/2}\int_{-h/2}^{h/2} u(rh + \xi, sh + \eta)^2 \, d\xi d\eta - h^2[Mu(rh, sh)]^2$$

$$\leqq \frac{h^2}{\pi^2}\int_{-h/2}^{h/2}\int_{-h/2}^{h/2} |\text{grad } u|^2 \, d\xi \, d\eta.$$

We sum this inequality over all the mesh points (rh, sh) for which S_{rs} intersects D and recall the definition (3.5) to find that

$$\mathscr{B}_h(Mu, Mu) \geqq \mathscr{B}(u, u) - \frac{h^2}{\pi^2}\mathscr{A}(u, u).$$

If $h^2 < \pi^2 \mu_n$, we see from this inequality and (3.10) that when $\mathscr{A}_1 = \mathscr{A}$, $\mathscr{B}_1 = \mathscr{B}$, $\mathscr{A}_2 = \mathscr{A}_h$ and $\mathscr{B}_2 = \mathscr{B}_h$, the inequalities (6.1), Chap. 3, and (6.5), Chap. 3, with $f(\xi) = g(\xi) = \xi - \pi^{-2}h^2$, are satisfied when u is any linear combination of the eigenvectors $u_1, \cdots, u_n$. Thus, the mapping principle shows that

$$\tilde{\mu}_n^{(h)} \geqq \mu_n - h^2/\pi^2,$$

provided the right-hand side is positive. Hence we have the upper bound

$$\mu_n \leqq \tilde{\mu}_n^{(h)} + h^2/\pi^2. \tag{3.14}$$

We have proved this inequality for $\mu_n > h^2/\pi^2$. Since $\tilde{\mu}_n^{(h)} \geqq 0$, it is still valid for $\mu_n \leqq h^2/\pi^2$.

For the more usual eigenvalue $\lambda_n = 1/\mu_n$, (3.14) gives the lower bound

$$\lambda_n \geqq \frac{\tilde{\lambda}_n^{(h)}}{1 + (h^2/\pi^2)\tilde{\lambda}_n^{(h)}}. \tag{3.15}$$

We remark that for the lowest eigenvalue λ_1, the author (1956) has obtained the better inequality

$$\lambda_1 \geqq \tilde{\lambda}_1^{(h)}$$

by using a nonlinear mapping M (see Chap. 7, Lemma 2.1) and some improvements by using recursion techniques. Hersch (1955) obtained the better bound

$$\lambda_1 > \frac{8}{\pi^2}\left(\sin^{-1}\sqrt{\frac{h^2\tilde{\lambda}_1^{(h)}}{8}}\right)^2$$

by using a different approach. These results do not appear to be capable of extension to higher eigenvalues.

Bounds of the form (3.15) have been obtained by the author (1958) for problems involving second order or even higher order operators with Dirichlet boundary conditions when the matrices A_{ij} and B_{ij} are formed in a specific manner.

We note that the bounds (3.8) and (3.15) involve the eigenvalues $\lambda_n^{(h)}$ and $\tilde{\lambda}_n^{(h)}$ of eigenvalue problems on two different mesh domains. Solving both of these problems doubles the necessary computational work. This situation can be remedied in several ways.

If D is convex and contains a disc of radius r, a dilatation of D by a factor $(1 + 3h/r)$ will take a mesh domain suitable for the upper bound (3.8) into a mesh domain suitable for the lower bound (3.15). Such a dilatation makes $\tilde{\lambda}_n^{(h)} = \lambda_n^{(h)}/(1 + 3h/r)^2$, so that we can get $\lambda_n^{(h)}$ from $\tilde{\lambda}_n^{(h)}$ or vice versa.

A second possibility is to modify the definition of M near the boundary so that the same mesh domain can be used for both problems. This idea has been carried through for various boundary conditions by Hubbard (1961), (1962). A simpler modification when D is a union of mesh squares was found by Hersch (1963). Other bounds of the kind given in this section have been obtained by Kuttler (1970b).

The definition (3.9) of the operator M is only one possibility, and it may be far from optimal. The work of Nitsche and Nitsche (1960) on inhomogeneous equations would seem to indicate that at least in some cases it is better to take averages over larger squares whose sides are proportional to $h^{2/5}$.

We observe in this connection that the eigenvalues μ_n of a given problem and the eigenvalues $\mu_n^{(h)}$ of a certain finite difference approximation to it have nothing to do with the mappings M and M_h used in obtaining either upper or lower bounds. It is an open problem to determine M_h and M which produce optimal bounds.

CHAPTER 7

Some Other Bounds

1. Some bounds for membrane problems. We consider the special case $k = l = 1$ of the inequality (9.6) of Chap. 3. It states that if $\mathscr{B}(u, u)$ is positive definite, if

$$\mathscr{A}(u, u) = \mathscr{A}_1(u, u) + \mathscr{A}_2(u, u)$$

and if the eigenvalues of the Rayleigh quotients $\mathscr{A}(u, u)/\mathscr{B}(u, u)$, $\mathscr{A}_1(u, u)/\mathscr{B}(u, u)$, and $\mathscr{A}_2(u, u)/\mathscr{B}(u, u)$ are $\lambda_1 \leqq \lambda_2 \leqq \cdots$, $\lambda_1^{(1)} \leqq \lambda_2^{(1)} \leqq \cdots$, and $\lambda_1^{(2)} \leqq \lambda_2^{(2)} \leqq \cdots$ respectively, then

$$\lambda_1 \geqq \lambda_1^{(1)} + \lambda_1^{(2)}. \tag{1.1}$$

We shall consider two applications of this inequality. Consider the two-dimensional problem

$$\begin{aligned} -\Delta u &= \lambda u \quad \text{in } D, \\ u &= 0 \quad \text{on } \partial D. \end{aligned} \tag{1.2}$$

We take

$$\mathscr{B}(u, u) = \int_D \int u^2 \, dx \, dy$$

and

$$\mathscr{A}(u, u) = \int_D \int \left[\left(\frac{\partial u}{\partial x} \right)^2 + \left(\frac{\partial u}{\partial y} \right)^2 \right] dx \, dy.$$

V is the space of twice continuously differentiable functions in D which vanish on the boundary. Let

$$\mathscr{A}_1(u, u) = \int_D \int \left(\frac{\partial u}{\partial x} \right)^2 dx \, dy,$$

$$\mathscr{A}_2(u, u) = \int_D \int \left(\frac{\partial u}{\partial y} \right)^2 dx \, dy.$$

We consider the problem of finding

$$\lambda_1^{(1)} = \inf_{u \in V} \frac{\int_D \int (\partial u/\partial x)^2 \, dx \, dy}{\int_D \int u^2 \, dx \, dy}.$$

If we go back to the proof of the existence of eigenvalues in § 1 of Chap. 3 we find that we should consider this problem on the completion of V with respect to the norm

$$\left\{\int_D\int\left[\left(\frac{\partial u}{\partial x}\right)^2 + u^2\right] dx\, dy\right\}^{1/2}.$$

Functions in this completion have no continuity properties in the variable y. Thus, the problem for $\lambda_1^{(1)}$ reduces the membrane to an infinite collection of parallel strings (see Hersch (1959)).

We consider the following one-dimensional problem:

$$\kappa_1 = \inf_{w(a)=w(b)=0} \frac{\int_a^b (dw/dx)^2\, dx}{\int_a^b w^2\, dx}. \tag{1.3}$$

This minimum problem gives the lowest eigenvalue of the one-dimensional problem

$$-d^2w/dx^2 = \kappa w, \qquad w(a) = w(b) = 0.$$

We easily see that $\kappa_1 = \pi^2/(b - a)^2$ with the eigenfunction $\sin \pi(x - a)/(b - a)$. By the definition (1.3) we see that if $w(a) = w(b) = 0$, then

$$\int_a^b \left(\frac{dw}{dx}\right)^2 dx \geqq \frac{\pi^2}{(b - a)^2}\int_a^b w^2\, dx.$$

The intersection of a line y = const. with D consists of one or more line segments $y = y_0, a_i(y_0) < x < b_i(y_0)$. Let L_x be a bound for the length of any such line segment produced by intersecting D with a line y = const. Then for any u in V,

$$\int_{a_i(y)}^{b_i(y)} \left[\frac{\partial u}{\partial x}(x, y)\right]^2 dx \geqq \frac{\pi^2}{(b_i - a_i)^2}\int_{a_i(y_0)}^{b_i(y_0)} u(x, y)^2\, dx$$

$$\geqq \frac{\pi^2}{L_x^2}\int_{a_i(y)}^{b_i(y)} u(x, y)^2\, dx.$$

We sum this inequality over all the segments corresponding to each y and integrate with respect to y to find that

$$\mathscr{A}_1(u, u) \geqq \frac{\pi^2}{L_x^2}\mathscr{B}(u, u).$$

Therefore,

$$\lambda_1^{(1)} \geqq \pi^2/L_x^2.$$

Similarly, if L_y is a bound for the line segments produced by intersecting D with lines x = const., we have

$$\lambda_1^{(2)} \geqq \pi^2/L_y^2.$$

Then (1.1) gives the lower bound

$$\lambda_1 \geqq \frac{\pi^2}{L_x^2} + \frac{\pi^2}{L_y^2} \tag{1.4}$$

for the lowest eigenvalue of the problem (1.2).

We may, of course, rotate coordinates to maximize this lower bound.

Example 1.1. Let D be as in Fig. 1.1.

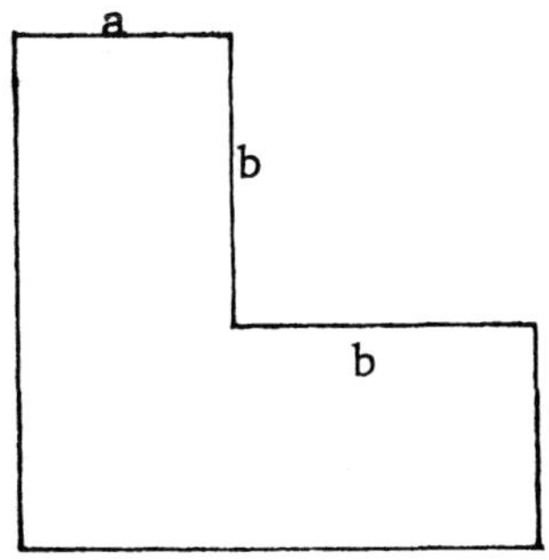

FIG. 1.1.

If the axes are horizontal and vertical we have $L_x = L_y = a + b$. The resulting lower bound is

$$\lambda_1 \geqq 2\pi^2/(a + b)^2.$$

If, instead, we take the axes at 45° to the horizontal and vertical, we find that $L_x = a\sqrt{2}$, $L_y = \min\{2a\sqrt{2}, (a + b)\sqrt{2}\}$. Then by (1.4),

$$\lambda_1 \geqq \frac{\pi^2}{2a^2} + \frac{\pi^2}{8} \max\left\{\frac{1}{a^2}, \frac{4}{(a + b)^2}\right\},$$

which is better for $b > (\sqrt{3} - 1)a$.

This result and analogous results for the more general problem

$$\begin{aligned} -\Delta u &= \lambda u \quad \text{in } D, \\ \partial u/\partial u + \alpha u &= 0 \quad \text{on } \partial D, \end{aligned} \tag{1.5}$$

as well as for vibrating plate problems have been found by Payne and the author (1957).

Another class of applications of the inequality (1.1) was obtained by Pólya and Schiffer (1953). Let $\mathscr{A}(u, u; \alpha)$ be a one-parameter family of positive definite quadratic functionals. Suppose that $\mathscr{A}(u, u; \alpha)$ is defined and convex in α on an interval. That is, for each u in V and any α_1 and α_2 in the interval

$$\mathscr{A}(u, u; \tfrac{1}{2}(\alpha_1 + \alpha_2)) \geqq \tfrac{1}{2}[\mathscr{A}(u, u; \alpha_1) + \tfrac{1}{2}\mathscr{A}(u, u; \alpha_2)]. \tag{1.6}$$

We again suppose that $\mathscr{B}(u, u)$ is positive definite and define

$$\lambda_1(\alpha) = \inf_{u \in V} \frac{\mathscr{A}(u, u; \alpha)}{\mathscr{B}(u, u)}.$$

If we put

$$\mathscr{A}(u, u) = \mathscr{A}(u, u; \tfrac{1}{2}(\alpha_1 + \alpha_2)),$$
$$\mathscr{A}_1(u, u) = \tfrac{1}{2}\mathscr{A}(u, u; \alpha_1),$$
$$\mathscr{A}_2(u, u) = \tfrac{1}{2}\mathscr{A}(u, u; \alpha_2),$$

we see from (1.1), (1.6) and the second monotonocity principle that

(1.7) $$\lambda_1(\tfrac{1}{2}(\alpha_1 + \alpha_2)) \geqq \tfrac{1}{2}[\lambda_1(\alpha_1) + \lambda_1(\alpha_2)].$$

That is, $\lambda_1(\alpha)$ is a convex function of α.

Example 1.2. For the problem (1.5) we have

$$\mathscr{A}(u, u; \alpha) = \int_D\int |\text{grad } u|^2 \, dx \, dy + \alpha \oint_{\partial D} u^2 \, ds,$$

$$\mathscr{B}(u, u) = \int_D\int u^2 \, dx \, dy.$$

$\mathscr{A}(u, u; \alpha)$ is clearly convex. Hence the lowest eigenfunction $\lambda_1(\alpha)$ is convex.

It is easily seen that $\lambda_1(0) = 0$ with the eigenfunction $u_1 = 1$.

We note that for any α

$$\mathscr{A}(u_1(x, y; \alpha), u_1(x, y; \alpha); \alpha) - \lambda_1(\alpha)\mathscr{B}((u_1(x, y; \alpha), u_1(x, y; \alpha)) = 0.$$

We formally differentiate with respect to α:

$$2\left[\mathscr{A}\left(u_1, \frac{\partial u_1}{\partial \alpha}\right) - \lambda_1(\alpha)\mathscr{B}\left(u_1, \frac{\partial u_1}{\partial \alpha}\right)\right] + \frac{\partial}{\partial \alpha}\mathscr{A}(u_1, u_1; \alpha) - \frac{d\lambda_1}{d\alpha}\mathscr{B}(u_1, u_1) = 0.$$

Since $\partial u_1/\partial \alpha$ is again in $V = H^1(D)$, the terms in the bracket cancel. Thus we find the expression

$$\frac{d\lambda_1}{d\alpha} = \frac{(\partial/\partial\alpha)\mathscr{A}(u_1, u_1; \alpha)}{\mathscr{B}(u_1, u_1)}.$$

(These steps can be justified.)

In Example 1.2, we have seen that at $\alpha = 0$ $\lambda_1 = 0$ and $u_1 = 1$. Hence,

$$\frac{d\lambda_1}{d\alpha}(0) = \frac{\oint_{\partial D} 1 \, ds}{\int_D\int 1 \, dx \, dy} = \frac{L}{A},$$

where L is the perimeter of D and A is its area.

The convexity of $\lambda_1(\alpha)$ now shows that for $\alpha > 0$ the upper bound

$$\lambda_1 \leqq \alpha L/A$$

holds. For $\alpha < 0$ we have the lower bound

$$\lambda_1 \geqq \alpha L/A.$$

Pólya and Schiffer have generalized the above convexity result considerably by showing that both the inequality (1.1) and the convexity (1.7) also hold for the sum $\lambda_1 + \lambda_2 + \cdots + \lambda_l$ of the l lowest eigenvalues, where l is any positive integer.

2. Symmetrization. The linearity of the mapping M in the mapping principle of § 6, Chap. 3, was needed in order to apply the Poincaré principle and thus obtain inequalities for eigenvalues beyond the first one. It is not needed to obtain an inequality for the first eigenvalue. We shall replace the mapping principle with the following simple lemma.

LEMMA 2.1. *Let $V_1, V_2, \mathscr{A}_1, \mathscr{B}_1, \mathscr{A}_2$, and $\mathscr{B}_2$ be as in the mapping principle of § 6, Chap. 3. Let the (not necessarily linear) transformation N map a subset S_1 of nonzero elements of V_1 which contains the eigenvector u_1 into nonzero elements of V_2. If*

$$\frac{\mathscr{B}_2(Nu, Nu)}{\mathscr{A}_2(Nu, Nu)} \geqq \frac{\mathscr{B}_1(u, u)}{\mathscr{A}_1(u, u)} \tag{2.1}$$

for all u in S_1, then

$$\mu_1^{(2)} \geqq \mu_1^{(1)}. \tag{2.2}$$

Proof. By the definitions of u_1 and $\mu_1^{(1)}$,

$$\mu_1^{(1)} = \frac{\mathscr{B}_1(u_1, u_1)}{\mathscr{A}_1(u_1, u_1)} \leqq \frac{\mathscr{B}_2(Nu_1, Nu_1)}{\mathscr{A}_2(Nu_1, Nu_1)} \leqq \mu_1^{(2)}.$$

We shall apply this lemma to a class of mappings known as symmetrizations. We consider the N-dimensional membrane problem

$$\begin{aligned} -\Delta u &= \lambda u \quad \text{in } D, \\ u &= 0 \quad \text{on } \partial D. \end{aligned} \tag{2.3}$$

We let

$$\begin{aligned} \mathscr{A}_1(u, u) &= \int_D |\text{grad } u|^2 \, dx, \\ \mathscr{B}_1(u, u) &= \int_D |u|^2 \, dx. \end{aligned} \tag{2.4}$$

We first show that the eigenfunction u_1 which corresponds to the lowest eigenvalue λ_1 of (2.3) does not vanish in D. To see this, we observe that the function $|u_1|$ also minimizes the Rayleigh quotient. By Theorem 2.1 of Chap. 3, it is a (strong) solution of the problem (2.3). Then because of the regularity theory for elliptic equations (see Fichera (1965)), $|u_1|$ is smooth. Since $-|u_1| \leqq 0$, the generalized maximum principle (Protter and Weinberger (1967, Thm. 2.10)) applied to a small neighborhood of each point of D shows that $|u_1| > 0$ in D. By multiplying by -1 if necessary, we shall make $u_1 > 0$ in D. Of course $u_1 = 0$ on the boundary.

Thus we must only define the mapping N for functions which are positive in D and vanish on the boundary. In order to define N we first define the symmetrization of a region about a hyperplane, say $x_1 = 0$.

DEFINITION (*Steiner symmetrization*). Let R be a closed bounded set in Euclidean K-space with smooth boundary. We say that R^* is obtained from R by *symmetrization about* $x_1 = 0$ if the intersection of R^* with each line L perpendicular to $x_1 = 0$ is a single line segment which is symmetric about $x_1 = 0$ and whose length is equal to the sum of the lengths of all the intersections of R with L.

Clearly R^* is symmetric about $x_1 = 0$. We can think of it as having been obtained from R by pushing the intersection of R with each L over to make it symmetric about $x_1 = 0$.

The following lemma is obvious.

LEMMA 2.2. *If the function* $f(x_2, \cdots, x_k)$ *is independent of* x_1, *and if* R^* *is obtained from* R *by symmetrization about* $x_1 = 0$, *then*

$$\int_{R^*} f dx = \int_R f dx.$$

Proof. We must only write the integrals as iterated integrals and observe that for fixed $x_2, \cdots, x_K$ the x_1-integral is the length of the intersection with the corresponding line L.

The next lemma is basic.

LEMMA 2.3. *If* R^* *is obtained from* R *by symmetrization about* $x_1 = 0$, *the surface area of the boundary of* R^* *is no greater than that of* R.

Proof. Assume that R can be represented in the form

$$R = \bigcup_{j=1}^{m} \{(x_1, x_2, \cdots, x_K) | \psi_{2j-1}(x_2, \cdots, x_K) \leqq x_1 \leqq \psi_{2j}(x_2, \cdots, x_K)\},$$

where the ψ_i are continuously differentiable for almost all $(x_2, \cdots, x_K)$ and continuous for all $(x_2, \cdots, x_K)$. (If the boundary of R contains line segments perpendicular to $x_1 = 0$, we approximate R by a set where this is not so.) The number m may depend on $(x_2, \cdots, x_K)$.

Then R^* can be represented as

$$R^* = \{(x_1, x_2, \cdots, x_K) | \, |x_1| \leqq \phi(x_2, \cdots, x_K)\},$$

where

$$\phi(x_2, \cdots, x_K) = \frac{1}{2} \sum_{j=1}^{m} [\psi_{2j} - \psi_{2j-1}]. \tag{2.5}$$

We can write the surface area S^* of R^* as

$$S^* = 2 \int \sqrt{1 + |\text{grad } \phi|^2} \, dx_2 \cdots dx_K, \tag{2.6}$$

where the integral is taken over the intersection of R^* with $x_1 = 0$ which is also the orthogonal projection of R on this plane. The gradient is in $k - 1$ space.

By (2.5) the K-vector $\{1, \text{grad}\,\phi\}$ satisfies the relation

$$\{1, \text{grad}\,\phi\} = \sum_{l=1}^{2m} \left\{\frac{1}{2m}, \frac{1}{2}(-1)^l \text{grad}\,\psi_l\right\}.$$

Hence by the triangle inequality

$$\sqrt{1 + |\text{grad}\,\phi|^2} \leqq \frac{1}{2}\sum_{l=1}^{2m} \sqrt{m^{-2} + |\text{grad}\,\psi_l|^2}$$

$$\leqq \frac{1}{2}\sum_{l=1}^{2m} \sqrt{1 + |\text{grad}\,\psi_l|^2}.$$

Substituting this in (2.6), we find that

$$S^* \leqq \sum_{l=1}^{2m} \int \sqrt{1 + |\text{grad}\,\psi_l|^2}\, dx_2 \cdots dx_K = S,$$

where S is the surface area of R. This completes the proof.

We now define the nonlinear transformation N. Let u be a function which is positive in D and vanishes on the boundary. Let R be the set

$$\{(x_1, \cdots, x_{N+1}) | (x_1, \cdots, x_N) \in D, 0 \leqq x_{N+1} \leqq u(x_1, \cdots, x_N)\}.$$

Let R^* be obtained by symmetrization of R in $x_1 = 0$. Then R^* has the representation

$$\{(x_1, \cdots, x_{N+1}) | (x_1, x_2, \cdots, x_N) \in D^*, 0 \leqq x_{N+1} \leqq u^*(x_1, \cdots, x_N)\},$$

where D^* is obtained from D by symmetrization in $x_1 = 0$. We define $N[u] = u^*$.

Thus N maps positive functions in D which vanish on the boundary into positive functions in D^* which vanish on the boundary.

We observe that

$$\mathscr{B}_1(u, u) = \int_D u^2 dx_1 \cdots dx_n = 2\int_R x_{N+1} dx_1 \cdots dx_{N+1}.$$

Hence if we define

$$\mathscr{B}_2(u, u) = \int_{D^*} u^2\, dx = 2\int_{R^*} x_{N+1} dx_1 \cdots dx_{N+1},$$

we see from Lemma 2.2 that

$$\mathscr{B}_2(Nu, Nu) = \mathscr{B}_1(u, u).$$

We now define R_ε to be the set $(x_1, \cdots, x_N) \in D, 0 \leqq x_{N+1} \leqq \varepsilon u(x_1, \cdots, x_N)$. If we symmetrize it, we obtain the corresponding set $R_\varepsilon^* : (x_1, \cdots, x_N) \in D^*$, $0 \leqq x_{N+1} \leqq \varepsilon u^*(x_1, \cdots, x_N)$. The areas of the bottom surfaces $x_{N+1} = 0$ of R_ε and R_ε^* are just the areas of D and D^*, which are equal by Lemma 2.2. Thus

Lemma 2.3 applied to R_ε tells us that

$$\int_D \sqrt{1 + \varepsilon^2 |\text{grad } u|^2}\, dx \geqq \int_{D^*} \sqrt{1 + \varepsilon^2 |\text{grad } u^*|^2}\, dx.$$

Subtracting the equal areas of D and D^* from both sides, we find that for small ε,

$$\text{(2.7)} \qquad \frac{1}{2}\varepsilon^2 \int_D |\text{grad } u|^2 dx + O(\varepsilon^4) \geqq \frac{1}{2}\varepsilon^2 \int_{D^*} |\text{grad } u^*|^2 dx + O(\varepsilon^4).$$

Dividing by $\frac{1}{2}\varepsilon^2$ and letting ε approach zero, we see that if we define

$$\mathscr{A}_2(u, u) = \int_{D^*} |\text{grad } u|^2 dx,$$

then[2]

$$\mathscr{A}_1(u, u) \geqq \mathscr{A}_2(Nu, Nu).$$

Thus we have the inequality (2.1). We see from Lemma 2.1 that if λ_1 is the lowest eigenvalue of the problem (3.3) and λ_1^* is the lowest eigenvalue of the corresponding problem for the symmetrized domain D^*, then

$$\lambda_1^* \leqq \lambda_1.$$

That is, *symmetrization of D about a (hyper)plane does not increase the lowest eigenvalue of the membrane problem* (2.3).

More drastic symmetrizations give similar results. We may symmetrize R about the set S: $x_1 = x_2 = \cdots = x_L = 0$ by replacing the intersection of each perpendicular hyperplane $(x_{L+1}, \cdots, x_K)$ constant with R by an L-dimensional ball of equal L-volume centered on S.

A similar argument shows that such a symmetrization applied to D does not increase λ_1. If $L = N$, we find the theorem of Faber (1923) and Krahn (1924): *Among all domains of given N-volume, the ball has the lowest value of λ_1 for the membrane problem* (2.3).

The proof of the fact that the Dirichlet integral does not increase under symmetrization depends very much on the fact that the terms of order lower than ε^2 cancel to give the inequality (2.7). If u does not vanish on the boundary, the boundary of R_ε has side walls of area

$$\varepsilon \int_{\partial D} u\, dS.$$

Therefore, there are noncanceling terms of order ε, and we cannot conclude that the Dirichlet integral decreases. The technique also fails if the function u changes sign.

[2] The idea of applying Steiner symmetrization to obtain such inequalities was introduced by Pólya and Szegö (1945). The above argument is due to Pólya (1948).

A different symmetrization which allows u to be nonzero on part of the boundary and which gives a lower bound for λ_1 under some additional assumptions has been defined by Bandle (1971b). Bandle (1971a) has also given an extension of symmetrization which permits one to obtain inequalities for problems with variable coefficients.

An excellent summary of the techniques of symmetrization can be found in the book of Pólya and Szegö (1951).

3. Non-self-adjoint problems. If the Hypotheses 1.1 of Chap. 3 are not satisfied, one is no longer assured that the problem

$$Au = \lambda Bu$$

has eigenvalues. The variational characterization of the eigenvalues disappears. However, in certain cases of differential operators some of the techniques which were developed for applications to variational problems can still be used.

If A is an elliptic differential operator of order $2M$ and if we let

$$(u, v) = \int_D u\bar{v}dx$$

it is often true that the quadratic functional

$$\mathrm{Re}\,(u, Au) \equiv \tfrac{1}{2}[(u, Au) + \overline{(u, Au)}]$$

is positive definite and coercive in the sense that

$$\|u\|^2_{H^M(D)} \leqq c\,\mathrm{Re}\,(u, Au)$$

for some constant c. The quadratic form $\mathrm{Re}\,(u, Au)$ does not correspond to the sesquilinear functional (u, Av), which is not Hermitian. However, it is easily seen that there is an elliptic operator A_0 of order $2M$ so that

$$(u, A_0 v) = \overline{(v, A_0 u)}$$

for all u, v in V and

$$(u, A_0 u) = \mathrm{Re}\,(u, Au).$$

Moreover, the operator $A - A_0$ is of lower order and hence can be considered as a perturbation of A_0.

Osborn (1966), (1967), (1971) assumed that the eigenfunctions of the self-adjoint problem

$$A_0 u = \lambda u$$

on V are known and used them in a Rayleigh–Ritz scheme. That is, he considered the eigenvalues of the matrix $(u_i^{(0)}, Au_j^{(0)})$, $i, j = 1, \cdots, k$, as approximations to the eigenvalues of the problem

$$Au = \lambda u$$

on V. He obtained various conditions under which the approximation error could be bounded by arbitrarily small bounds.

Vainikko (1964), (1967) obtained error bounds in terms of approximability estimates when the eigenvalues are computed in terms of finite matrices of the form (v_i, Aw_j) with more or less arbitrary elements v_i and w_j.

Bramble and Osborn (1972), (1973) have obtained results of the same type in the setting of finite element methods.

References

N. ARONSZAJN (1948), *The Rayleigh–Ritz and A. Weinstein methods for approximation of eigenvalues. I: Operators in a Hilbert space. II: Differential operators*, Proc. Nat. Acad. Sci. U.S.A., 34, pp. 474–480, 594–601.

——— (1949–1950), *The Rayleigh–Ritz and the Weinstein methods for approximation of eigenvalues*, Tech. Reps. 1–4, Oklahoma A & M College, Stillwater.

——— (1951), *Approximation methods for eigenvalues of completely continuous symmetric operators*, Proc. Symposium on Spectral Theory and Differential Problems, Oklahoma A & M College, Stillwater, pp. 179–202.

N. ARONSZAJN AND A. WEINSTEIN (1942), *On the unified theory of eigenvalues of plates and membranes*, Amer. J. Math., 64., pp. 623–645.

J. P. AUBIN (1967), *Behavior of the error of the approximate solution of boundary value problems for linear elliptic operators by Galerkin's and finite difference methods*, Ann. Scuola Norm. Sup. Pisa (3), 21, pp. 599–637.

——— (1968), *Evaluation des erreurs de troncature des approximations des espaces de Sobolev*, J. Math. Anal. Appl., 21, pp. 356–368.

J. P. AUBIN AND H. G. BURCHARD (1971), *Some aspects of the method of the hypercircle applied to elliptic variational problems*, Numerical Solution of Partial Differential Equations, vol. II, B. Hubbard, ed., Academic Press, New York, pp. 1–67.

I. BABUSKA (1971), *The finite element method for elliptic differential equations*, Numerical Solution of Partial Differential Equations, vol. II, B. Hubbard, ed., Academic Press, New York, pp. 69–106.

C. BANDLE (1971a), *Konstruktion isoperimetrischer Ungleichungen der mathematischen Physik aus solchen der Geometrie*, Comment. Math. Helv., 46, pp. 182–213.

——— (1971b), *Extremaleigenschaften von Kreissektoren und Halbkugeln*, Ibid., 46, pp. 356–380.

L. BAUER AND E. REISS (1965), *Nonlinear buckling of rectangular plates*, SIAM J. Appl. Math., 13, pp. 603–626.

N. W. BAZLEY (1959), *Lower bounds for eigenvalues with applications to the helium atom*, Proc. Nat. Acad. Sci. U.S.A., 45, pp. 850–853.

——— (1961), *Lower bounds for eigenvalues*, J. Math. Mech., 10, pp. 289–308.

N. W. BAZLEY AND D. W. FOX (1961), *Truncation in the method of intermediate problems for lower bounds to eigenvalues*, J. Res. Nat. Bur. of Standards, 65B, pp. 105–111.

——— (1961), *Lower bounds for eigenvalues of Schrödinger's equation*, Phys. Rev., 124, pp. 483–492.

——— (1963), *Lower bounds for energy levels of molecular systems*, J. of Mathematical Phys., 4, pp. 1147–1153.

M. BERGER AND P. FIFE (1968), *On von Kármán's equations and the buckling of a thin elastic plate, II*, Comm. Pure Appl. Math., 21, pp. 227–241.

G. BERTRAM (1957), *Fehlerabschätzung fur das Ritz–Galerkinsche Verfahren bei Eigenwertproblemen*, Z. Angew. Math. Mech., 37, pp. 191–201.

G. BIRKHOFF (1969), *Piecewise bicubic interpolation and approximation in polygons*, Approximation with Special Emphasis on Spline Functions, I. J. Schoenberg, ed., Academic Press, New York, pp. 185–221.

G. BIRKHOFF AND C. DEBOOR (1965), *Piecewise polynomial interpolation and approximation*, Approximation of Functions, H. L. Garabedian, ed., Elsevier, New York, pp. 164–190.

G. BIRKHOFF, G. DEBOOR, B. SWARTZ AND B. WENDROFF (1966), *Rayleigh-Ritz approximation by piecewise polynomials*, SIAM J. Numer. Anal., 3, pp. 188–203.

G. BIRKHOFF AND G. FIX (1970), *Accurate eigenvalue computations for elliptic problems*, Numerical Solution of Field Problems in Continuum Physics, vol. II, G. Birkhoff and R. S. Varga, eds., SIAM-AMS Proceedings, American Mathematical Society, Providence, R. I., pp. 111–151.

G. BIRKHOFF, M. H. SCHULTZ AND R. S. VARGA (1968), *Piecewise Hermite interpolation in one and two variables with applications to partial differential equations*, Numer. Math., 11, pp. 232–256.

W. BÖRSCH-SUPAN (1958), *Obere Schranken für den gröszten Eigenwert eines vollstetigen selbstadjungierten Operators*, Math. Ann., 134, pp. 453–457.

J. H. BRAMBLE AND S. R. HILBERT (1970), *Estimation of linear functionals on Sobolev spaces with applications to Fourier transforms and spline interpolation*, SIAM J. Numer. Anal., 7, pp. 112–124.

——— (1971), *Bounds for a class of linear functionals with applications to Hermite interpolation*, Numer. Math., 16, pp. 362–369.

J. H. BRAMBLE AND J. E. OSBORN (1972), *Approximation of Steklov eigenvalues of non-selfadjoint second order elliptic operators*, The Mathematical Foundations of the Finite Element Method, A. K. Aziz, ed., Academic Press, New York.

——— (1973), *Rate of convergence estimates for non-selfadjoint eigenvalue approximations*, Math. Comp., 27, pp. 525–550.

J. H. BRAMBLE AND L. E. PAYNE (1963), *Upper and lower bounds in equations of forced vibration type*, Arch. Rational Mech. Anal., 14, pp. 153–170.

J. H. BRAMBLE AND M. ZLAMAL (1970), *Triangular elements in the finite element method*, Math. Comp., 24, pp. 809–819.

R. E. CARLSON AND C. A. HALL (1971), *On piecewise polynomial interpolation in rectangular polygons*, J. Approx. Theor., 4, pp. 37–53.

P. G. CIARLET, M. H. SCHULTZ AND R. S. VARGA (1967), *Numerical methods of high-order accuracy for nonlinear boundary value problems. I: One dimensional problem*, Numer. Math., 9, pp. 394–430.

——— (1968), *Numerical methods of higher order accuracy for nonlinear boundary value problems. III: Eigenvalue problems*, Ibid., 12, pp. 120–133.

L. COLLATZ (1937), *Konvergenzbeweis und Fehlerabschätzung für das Differenzenverfahren bei Eigenwertproblemen gewöhnlicher Differenzialgleichungen zweiter und vierter Ordnung*, Deutsche Math., 2, pp. 189–215.

——— (1940), *Schrittweise Näherungen bei Integralgleichungen und Eigenwertschranken*, Math. Z., 46, pp. 692–708.

——— (1942), *Einschlieszungssatz für die characteristischen Zahlen von Matrizen*, Ibid., 48, pp. 221–226.

——— (1948), *Eigenwertprobleme und ihre numerische Behandlung*, Chelsea, New York.

R. COURANT (1920), *Über die Eigenwerte bei den Differentialgleichungen der mathematischen Physik.*, Math. Z., 7, pp. 1–57.

——— (1927), *Über direkte Methoden in der Variationsrechnung und über verwandte Fragen*, Math. Ann., 97, pp. 711–736.

——— (1943), *Variational methods for the solution of problems of equilibrium and vibrations*, Bull. Amer. Math. Soc., 49, pp. 1–23.

C. DAVIS AND W. M. KAHAN (1970), *The rotation of eigenvectors by a perturbation. III*, SIAM J. Numer. Anal., 7, pp. 1–46.

L. DEVITO (1956), *Sugli autovalori e sulle autosoluzioni di una classe di trasformazioni hermitiane*, Rend. Sem. Mat. Univ. Padova, 25, pp. 144–175.

——— (1961), *Sul calcolo approssimato degli autovalori delle trasformazioni compatte e delle relative molteplicità*, Rend. Accad. Naz. Lincei, 30, pp. 351–356 and pp. 452–459.

J. B. DIAZ (1956), *Upper and lower bounds for eigenvalues*, Proc. Eighth Symposium on Applied Mathematics of the A.M.S. (Calculus of Variations and its Applications), American Mathematical Society, Providence, R.I., pp. 53–78.

G. FABER (1923), *Beweis, dass unter allen homogenen Membranen von gleicher Fläche und gleicher Spannung die kreisförmige den tiefsten Grundton gibt*, Bayr. Akad. Wiss. Math.-Natur. Kl. S.-B., pp. 169–172.

G. FICHERA (1965), *Linear elliptic differential systems and eigenvalue problems*, Lecture Notes in Math. no. 8, Springer, Berlin.

——— (1966), *Approximation and estimates for eigenvalues*, Numerical Solution of Partial Differential Equations. I. J. Bramble, ed., Academic Press, New York, pp. 317–352.

——— (1971), *Further developments in the approximation theory of eigenvalues*, Numerical Solution of Partial Differential Equations. II, B. Hubbard, ed., Academic Press, New York, pp. 243–252.

P. C. FIFE (1970), *The Bénard problem for general fluid dynamical equations and remarks on the Boussinesq approximation*, Indiana Univ. Math. J., 20, pp. 303–326.

G. FIX AND G. STRANG (1969), *Fourier analysis of the finite element method in Ritz–Galerkin theory*, Studies in Appl. Math., 48, pp. 265–273.

L. FOX, P. HENRICI AND C. MOLER (1967), *Approximations and bounds for eigenvalues of elliptic operators*, SIAM J. Numer. Anal., 4, pp. 89–103.

K. O. FRIEDRICHS (1948), *Criteria for the discrete character of the spectra of ordinary differential operators*, Courant Anniversary Volume, Interscience, New York, pp. 145–160.

M. GOLOMB AND H. F. WEINBERGER (1969), *Optimal approximation and error bounds*, On Numerical Approximation, R. Langer, ed., Univ. of Wisconsin Press, Madison, pp. 117–190.

S. H. GOULD (1966), *Variational Methods for Eigenvalue Problems: An Introduction to the Weinstein Method of Intermediate Problems*, 2nd ed., Univ. of Toronto Press, Toronto.

C. A. HALL (1969), *Bicubic interpolation over triangles*, J. Math. Mech., 19, pp. 1–11.

A. HAMMERSTEIN (1927), *Eine Restabschätzung für das Ritzsche Verfahren bei gewissen Variationsproblemen mit Nebenbedingungen*, S.-B. Berliner Math. Ges., 26, pp. 171–177.

J. HERSCH (1955), *Équations differentielles et fonctions des cellules*, C. R. Acad. Sci., Paris, 240, pp. 1602–1605.

——— (1959), *Une interprétation du principe de Thomson et son analogue pour la fréquence fondamentale d'une membrane. Application*, Ibid., 248, pp. 2060–2062.

——— (1963), *Lower bounds for all eigenvalues by cell functions: a refined form of H. F. Weinberger's method*, Arch. Rational Mech. Anal., 12, pp. 361–366.

B. E. HUBBARD (1961), *Bounds for eigenvalues of the free and fixed membrane by finite difference methods*, Pacific J. Math., 11, pp. 559–590.

——— (1962), *Eigenvalues of the non-homogeneous rectangular membrane by finite difference methods*, Arch. Rational Mech. Anal., 9, pp. 121–133.

V. I. IUDOVICH (1966), *On the origin of convection*, Prikl. Mat. Mekh., 30, pp. 1000–1005; J. Appl. Math. Mech., 30, pp. 1193–1199.

W. KAHAN (1967), *Inclusion theorems for clusters of eigenvalues of hermitian matrices*, Report, Dept. of Computer Science, Univ. of Toronto, Toronto.

T. KATO (1949), *On the upper and lower bounds for eigenvalues*, J. Phys. Soc. Japan, 4, pp. 334–339.

——— (1966), *Perturbation Theory for Linear Operators*, Springer, Berlin.

T. KATO, H. FUJITA, Y. NAKATA AND M. NEWMAN (1957), *Estimation of the frequencies of thin elastic plates with free edges*, J. Res. Nat. Bur. Standards, 59, pp. 169–186.

H. B. KELLER (1965), *On the accuracy of finite difference approximations to the eigenvalues of differential and integral operators*, Numer. Math., 7, pp. 412–419.

B. KNAUER (1971), *Untere Schranken für die Eigenwerte selbstadjungierter positivdefiniter Operatoren*, Ibid., 17, pp. 166–171.

G. KNIGHTLY AND D. SATHER (1970), *On nonuniqueness of solutions of the von Kármán equations*, Arch. Rational Mech. Anal., 36, pp. 65–78.

——— (1972), *Nonlinear buckled states of rectangular plates*, Summ. Rep. 1230, Mathematics Research Center, Madison.

F. KOEHLER (1957), *Estimates for the eigenvalues of infinite matrices*, Pacific J. Math., 7, pp. 1391–1404.

W. KOHN (1947), *A note on Weinstein's variational method*, Phys. Rev., 71, pp. 902–904.

E. KRAHN (1924), *Über eine von Rayleigh formulierte Minimaleigenschaft des Kreises*, Math. Ann., 94, pp. 97–100.

E. KREYSZIG (1954), *Die Einschliessung von Eigenwerten hermitescher Matrizen beim iterationsverfahren*, Z. Angew. Math. Mech., 34, pp. 459–469.

——— (1955), *Die Ausnützung zusätzlicher Vorkentnisse für die Einschliessung von Eigenwerten beim Iterationsverfahren*, Ibid., 35, pp. 89–95.

N. KRYLOV (1928), *Sur l'application du principe de minimum à la théorie des oscillations propres des systèmes*, Acta Math., 52, pp. 135–141.

——— (1931), *Les méthodes de solution approchée des problèmes de la physique mathématique*, Mémr. Sci. Math., no. 49, Gauthier-Villars, Paris.

N. KRYLOV AND N. BOGOLIUBOV (1929), *Sur le calcul des racines de la transcendante de Fredholm les plus voisines d'une nombre donné par les méthodes des moindres carrés et de l'algorithme variationel*, Isv. Akad. Nauk CCCP, Classe des Sci. Phisico Math., VII Serie Leningrad, pp. 471–488.

S. T. KURODA (1961), *On a generalization of the Weinstein–Aronszajn formula and the infinite determinant*, Sci. Papers College Gen. Educ. Univ. Tokyo, 11, pp. 1–12.

J. R. KUTTLER (1970a), *Finite difference approximations for eigenvalues of uniformly elliptic operators*, SIAM J. Numer. Anal., 7, pp. 206–232.

——— (1970b), *Upper and lower bounds for eigenvalues by finite differences*, Pacific J. Math., 35, pp. 429–440.

N. J. LEHMANN (1949)–(1950), *Beiträge zur numerischen Lösung linearer Eigenwertprobleme*, Z. Angew. Math. Mech., 29, pp. 342–356 and 30, pp. 1–16.

——— (1963), *Optimale Eigenwerteinschliessungen*, Numer. Math., 5, pp. 246–272.

H. J. MAEHLY (1952), *Ein neues Variationsverfahren zur genäherten Berechnung der Eigenwerte hermitescher Operatoren*, Helv. Phys. Acta, 25, pp. 547–568.

C. MOLER AND L. E. PAYNE (1968), *Bounds for eigenvalues and eigenvectors of symmetric operators*, SIAM J. Numer. Anal., 5, pp. 64–70.

J. A. NITSCHE (1959), *Einfache Fehlerschranken beim Eigenwertproblem symmetrischer Matrizen*, Z. Angew. Math. Mech., 39, pp. 322–325.

J. A. NITSCHE AND J. C. NITSCHE, *Error estimates for the numerical solution of elliptic differential equations*, Arch. Rational Mech. Anal., 5, pp. 293–306.

J. E. OSBORN (1966), *Approximation of the eigenvalues of nonselfadjoint operators*, J. Math. and Phys., 45, pp. 391–401.

———(1967), *Approximation of eigenvalues of a class of unbounded, nonselfadjoint operators*, SIAM J. Numer. Anal., 4, pp. 45–54.

——— (1971), *A method for approximating the eigenvalues of nonselfadjoint ordinary differential operators*, Mem. Accad. Sci. Torino, 4, no. 14.

M. PARODI (1952), *Sur quelques propriétés des valeurs caractéristiques des matrices carrées*, Mém. Sci. Math., no. 118, Gauthier-Villars, Paris.

L. E. PAYNE (1954), *Inequalities for eigenvalues of membranes and plates*, J. Rational Mech. Anal., 4, pp. 517–529.

——— (1967), *Isoperimetric inequalities and their applications*, SIAM Rev., 9, pp. 453–488.

L. E. PAYNE AND H. F. WEINBERGER (1957), *Lower bounds for vibration frequencies of elastically supported membranes and plates*, J. Soc. Indust. Appl. Math., 5, pp. 171–182.

J. G. PIERCE AND R. S. VARGA (1972a), *Higher order convergence results for the Rayleigh–Ritz method applied to eigenvalue problems; I: Estimates relating Rayleigh–Ritz and Galerkin approximations to eigenfunctions*, SIAM J. Numer. Anal., 9, pp. 137–151.

——— (1972b), *Higher order convergence results for the Rayleigh–Ritz method applied to eigenvalue problems. II: Improved error bounds for eigenfunctions*, Numer. Math., 19, pp. 155–169.

H. POINCARÉ (1890), *Sur les équations aux dérivées partielles de la physique mathématique*, Amer. J. Math., 12, pp. 211–294.

G. PÓLYA (1948), *Torsional rigidity, principal frequency, electrostatic capacity, and symmetrization*, Quart. Appl. Math., 6, pp. 267–277.

——— (1952), *Sur une interprétation de la méthode des différences finie qui peut fournir des bornes supérieures ou inférieures*, C. R. Acad. Sci. Paris, 235, pp. 995–997.

——— (1968), *Graeffe's method for eigenvalues*, Numer. Math., 11, pp. 315–319.

G. PÓLYA AND M. SCHIFFER (1953–54), *Convexity of functionals by transplantation*, J. Analyse Math., 3, pp. 245–345.

G. Pólya and G. Szegö (1945), *Inequalities for the capacity of a condenser*, Amer. J. Math., 67, pp. 1–32.

——— (1951), *Isoperimetric inequalities in mathematical physics*, Princeton Univ. Press, Princeton, N.J.

M. H. Protter and H. F. Weinberger (1967), *Maximum principles in differential equations*, Prentice-Hall, Englewood Cliffs, N.J.

P. Rabinowitz (1968), *Existence and nonuniqueness of rectangular solutions of the Bénard problem*, Arch. Rational Mech. Anal., 29, pp. 32–57.

Lord Rayleigh (J. W. Strutt) (1878), *The Theory of Sound*, Macmillan, London; 2nd ed., Macmillan, London, 1894; Dover, New York, 1945.

——— (1899), *On the calculation of the frequency of vibration of a system in its gravest mode, with an example from hydrodynamics*, Philos. Mag., 47, pp. 566–572. Scientific Papers, vol. 4, Dover, New York, 1964, pp. 407–412.

W. Ritz (1909), *Über eine neue Methode zur Lösung gewisser Variationsprobleme der mathematischen Physik*, J. Reine u. Angew. Math., 135, pp. 1–61.

——— (1909b), *Theorie der Transversalschwingungen einer quadratischen Platte mit freien Rändern*, Ann. Physik, 28, pp. 737–786.

E. Saibel (1951), *Free and forced vibrations of composite systems*, Proc. Symposium on Spectral Theory and Differential Problems, Oklahoma A & M College, pp. 333–343.

D. H. Sattinger (1973), *Topics in Stability and Bifurcation Theory*, Springer Lecture Notes, no. 309, Springer, Berlin, 1973.

V. K. Saulev (1955), *On the solution of the problem of eigenvalues by the method of finite differences*, Vychisl. Mat. Vychisl. Tehn., 2, pp. 116–144; English Transl. Amer. Math. Soc. Transl. (2), 8 (1958), pp. 257–287.

M. H. Schultz (1969a), *Multivariate spline functions and elliptic problems*, Approximation with Special Emphasis on Spline Functions, I. J. Schoenberg, ed., Academic Press, New York, pp. 279–347.

——— (1969b), *L^2-multivariate approximation theory*, SIAM J. Numer. Anal., 6, pp. 184–209.

——— (1969c), *Rayleigh–Ritz–Galerkin methods for multidimensional problems*, Ibid., 6, pp. 523–538.

——— (1969a), *Approximation theory of multivariate spline functions in Sobolev spaces*, Ibid., 6, pp. 570–582.

M. H. Schultz and R. S. Varga (1967), *L-splines*, Numer. Math., 10, pp. 345–369.

V. S. Sorokin (1953), *Variatsionnyi metod v teorii konvektsii*, Prikl. Mat. Mekh., 17, pp. 39–48.

G. Strang (1971), *The finite element method and approximation theory*, Numerical Solution of Partial Differential Equations, vol. II, B. Hubbard, ed., Academic Press, New York, pp. 547–583.

G. Strang and G. Fix (1973), *An Analysis of the Finite Element Method*, Prentice-Hall, Englewood Cliffs, N.J.

J. L. Synge (1957), *The Hypercircle in Mathematical Physics*, Cambridge Univ. Press, Cambridge.

O. Taussky (1954), *Contributions to the solution of systems of linear equations and the determination of eigenvalues*, U.S. Government Printing Office, Washington, D.C.

G. Temple (1928), *The computation of characteristic numbers and characteristic functions*, Proc. London Math. Soc., 29, pp. 257–280.

G. Temple and W. G. Bickley (1933), *Rayleigh's Principle and Its Applications to Engineering*, Constable, London.

E. Trefftz (1933), *Über Fehlerabschätzung bei Berechnung von Eigenwerten*, Math. Ann., 108, pp. 595–604.

M. R. Ukhovskii and V. I. Iudovich (1963), *On the equations of steady-state convection*, Prikl. Mat. Mekh., 27, pp. 295–300; Appl. Math. Mech., 27, pp. 432–440.

G. M. Vainikko (1964), *Asymptotic evaluation of the error of projection methods for the eigenvalue problem*, Zh. Vychisl. Mat. i Mat. Fiz., 4, pp. 405–425; U.S.S.R. Comput. Math. and Math. Phys., 4, pp. 9–36.

——— (1967), *On the speed of convergence of approximate methods in the eigenvalue problem*, Zh. Vychisl. Mat. i Mat. Fiz., 7, pp. 977–987; U.S.S.R. Comput. Math. and Math. Phys., 7, pp. 18–32.

R. S. VARGA (1971), *Functional Analysis and Approximation Theory in Numerical Analysis*, Regional Conference Series in Appl. Math. 3, Society for Industrial and Applied Mathematics, Philadelphia.

L. VEIDINGER (1965), *Evaluation of the error in finding eigenvalues by the method of finite differences*, Zh. Vychisl. Mat. i Mat. Fiz., 5, pp. 806–815; U.S.S.R. Comput. Math. and Math. Phys., 5, pp. 28–42.

A. G. WALKER AND J. D. WESTON (1949), *Inclusion theorems for the eigenvalues of a normal matrix*, J. London Math. Soc., 29, pp. 28–31.

K. WASHIZU (1952–53), *Geometrical representations of bounds of eigenvalues*, J. Japan Soc. for Appl. Mech., 30–31, pp. 29–32.

——— (1955), *On the bounds of eigenvalues*, Quart. J. Mech. Appl. Math., 8, pp. 311–325.

H. F. WEINBERGER (1952a), *Error estimation in the Weinstein method for eigenvalues*, Proc. Amer. Math. Soc., 3, pp. 643–646.

——— (1952b), *An optimum problem in the Weinstein method for eigenvalues*, Pacific J. Math., 2, pp. 413–418.

——— (1954), *A Rayleigh–Ritz procedure giving upper and lower bounds for eigenvalues*, Inst. for Fluid Dynamics and Applied Mathematics, Tech. Note BN-41, Univ. of Maryland, College Park.

——— (1955), *An extension of the classical Sturm–Liouville theory*, Duke Math. J., 22, pp. 1–14.

——— (1956), *Upper and lower bounds for eigenvalues by finite difference methods*, Comm. Pure Appl. Math., 9, pp. 613–623. (Also, Proc. Conference on Partial Differential Equations, Berkeley, Calif., 1955).

——— (1958), *Lower bounds for higher eigenvalues by finite difference methods*, Pacific J. Math., 8, pp. 339–368.

——— (1959), *A theory of lower bounds for eigenvalues*, Tech. Note BN-183, Inst. for Fluid Dynamics and Applied Mathematics, Univ. of Maryland, College Park.

——— (1960), *Error bounds in the Rayleigh–Ritz approximation of eigenvectors*, J. Res. Nat. Bur. Standards, 64B, pp. 217–225.

——— (1962), *Variational methods for eigenvalue problems*. Lecture Notes, Univ. of Minn. Engineering Book Store, Minneapolis.

A. WEINSTEIN (1935), *Sur la stabilité des plaques encastrées*, C. R. Acad. Sci. Paris, 200, pp. 107–109.

——— (1937), *Étude des spectres des équations aux dérivées partielles de la théorie des plaques élastiques*, Mémor. Sci. Math., no. 88, Gauthier-Villars, Paris.

——— (1951), *Quantitative methods in Sturm–Liouville theory*, Proc. Symposium on Spectral Theory and Differential Problems, Oklahoma A & M College, Stillwater, pp. 345–352.

——— (1953), *Variational methods for the approximation and exact computation of eigenvalues*, Nat. Bur. Standards Appl. Math. Ser., 29, pp. 83–89.

——— (1963), *On the Sturm–Liouville theory and the eigenvalues of intermediate problems*, Numer. Math., 5, pp. 238–245.

A. WEINSTEIN AND W. STENGER (1972), *Methods of Intermediate Problems for Eigenvalues*, Academic Press, New York.

D. H. WEINSTEIN (1934), *Modified Ritz method*, Proc. Nat. Acad. Sci. U.S.A., 20, pp. 529–532.

B. WENDROFF (1965), *Bounds for eigenvalues of some differential operators by the Rayleigh–Ritz method*, Math. Comp., 19, pp. 218–224.

H. WEYL (1912), *Das asymptotische Verteilungsgesetz der Eigenwerte linearer partieller Differenzialgleichungen (mit einer Anwendung auf die Theorie der Hohlraumstrahlung)*, Math. Ann., 71, pp. 441–479.

H. WIELANDT (1953), *Inclusion theorems for eigenvalues*, Nat. Bur. of Standards Appl. Math. Ser., 29, pp. 75–78.

——— (1954), *Einschliessung von Eigenwerten hermitescher Matrizen nach dem Abschnittsverfahren*, Arch. Math., 5, pp. 108–114.

M. ZLAMAL (1968), *On the finite element method*, Numer. Math., 12, pp. 394–409.

INDEX

Page numbers for definitions are in italics

A

Adjoint
  space, *25*, 27
  transformation, 118, 119
Aronszajn, N., 78, 84, 86, 91
Aronszajn's inequality, 91
  method, 84–88, 91, 96–98, 103, 121
  rule, 78
Aubin, J. P., 132

B

Babuska, I., 132
Banach space, *24*
Bandle, C., 149
Bauer, L., 17
Bazley, N. W., 88, 89, 91
Bénard problem, 17, 34
Berger, M., 17
Bertram, G., 113
Bessel's inequality, 51
Bickley, W. G., 115
Bifurcation, 8, 12, 13, 15, 17, 18
Bilinear functional (form), *22*, 23
Birkhoff, G., 70, 126, 132
Bogoliubov, N., 115
Börsch-Supan, W., 113
Boundary conditions, natural, stable, unstable, 59
Bounded linear functional, *25*, 25–29, 43–44, 54
Bramble, J. H., 126, 132, 150
Buckling, 12, 15, 17, 59
Burchard, H. G., 132

C

Carlson, R. E., 132
Cauchy criterion, *24*
Cauchy sequence, *24*
Ciarlet, P. G., 132
Collatz, L., 115
Compact, *53*
Complete linear vector space, *24*, 25
Completion, 26, 28, 29, 59–60, 61, 86
Complete continuity, *50*, 50–52, 53
Complexification, 28
Continuous spectrum, *44*, 45, 49, 50
Convergence, *24*, 71, 84, 86, 103
Courant, R., 56, 128, 135
Courant–Weyl principle, 56–57

D

Davis, C., 126
De Boor, C., 70, 126, 132
De Vito, L., 116
Diagonalization of matrices, 3, 11, 40
Dimension *m*, *21*
Discrete spectrum, *44*, 50
Dual space, *21*

E

Eigenfunction, *1*
  approximation of, 123–26
  positivity of first, 145
  Rayleigh–Ritz, 124
Eigenvalue(s), *1*, 1–6, 31, 36–41
  error bounds for in Rayleigh–Ritz method, 67–74
  existence of, 41
  formulation of self-adjoint, problem, 31–45
  matrix, 3, 10, 11, 40
  of Rayleigh quotient, *45*
  reality of, 37–38
  as successive maxima, 39
  of a sum of operators, 65
Eigenvectors, 36–39, 52
  error bounds for, 123–26
  expansion in terms of, 52
  orthogonality of, 38
  Rayleigh–Ritz, 124–25
Elliptic differential operator, 35, 54
Error bounds, 86, 94, 112, 132, 133, 150
  a priori and a posteriori, 67
  for eigenvalues in Rayleigh–Ritz method, 67–74, 111
  for eigenvectors, 123–26
Essential spectrum, *44*, 45, 81
Extension across the boundary, 73, 132

F

Faber, G., 148
Fichera, G., 37, 54, 73, 88, 116, 119–21, 130, 145
Fichera's method, 115–21
Fife, P. C., 17, 18
Finite difference equations, 127–39
  from Rayleigh–Ritz method, 127–29
  upper and lower bounds from, 133–39
Finite elements method, 129–33, 150
Fix, G., 132
Fourier series, 52
Fox, D. W., 89, 91
Fox, L., 115
Fredholm invariants, 121
Friedrichs, K. O., 41
Friedrichs' inequality, 46
Functional
  bilinear, *22*, 23
  linear, *21*, 25–29, 41–44, 54, 63
  sesquilinear, *22*, 23, 31, 35, 53–55

G

Gould, S. H., 84
Green's function, 117
  operator, 117, 118, 120

H

Hall, C. A., 132
Hammerstein, A., 113
Heat flow, problem of, 3, 32, 50, 59, 60, 61, 94, 105, 107, 118, 128, 134–39, 141–45, 148
Henrici, P., 115
Hermite interpolation functions, *131*
Hermitian
  functional, *22*, 23, 31, 35, 54
  kernel, 117
  matrix, *22*, 32, 40, 77
Hersch, J., 139, 142
Hilbert, S. R., 132
Hilbert space, *25*, 25–28, 86, 88
Hubbard, B. E., 139
Hypotheses, 31, 54, 106

I

Inclusion theorems, 114–15
Infinite-dimensional, *21*
Inhomogeneous equations, 3, 53–55
Iudovich, V. I., 18

K

Kahan, W. M., 115, 126
Kato, T., 81, 115
Knauer, B., 112
Knightly, G., 17
Koehler, F., 112
Kohn, W., 115
Krahn, E., 148
Kreyszig, E., 115
Krylov, N., 71, 115, 133, 136
  error bounding techniques of, 70, 71, 133
Kuroda, S. T., 81
Kuttler, J. R., 139

L

Lagrange equations, 9, 13
Lax, Peter, 57
Least upper bound, 38, 39
  sufficient condition for attainment of, 41
Lehmann, N. J., 114
Linear functional(s), *21*, 25–29, 41–44, 54, 63
  bounded, *25*, 25–29, 43–44, 54
Linearly dependent, *21*, 23
Linearly independent
  elements, *21*
  functionals, 41
Linear operator, *20*, 21
  self-adjoint, *37*
  completely continuous (compact), *53*
Linear subspace, *21*, 56
Linear transformation, *20*, 21
Linear vector spaces, *19*, 19–21
  normed, *23–24*, 24–25

M

Maehly, H. J., 114
Mapping principle, 57, 58, 62, 68, 70, 134–36, 138, 145
Matrix
  diagonalization, 3, 11, 40
  positive definite, *10*
  sparse, *128*, 129, 134
Matrix eigenvalues, 3, 10, 11, 40
  complementary bounds in terms of, 98–109
Membrane problems, bounds for, 32, 50, 59, 60, 61, 94, 105, 107, 118, 128, 134–39, 141–45, 148
Mesh function, 127–28, 134, 136
Mesh points, 127–28, 131
Moler, C., 115

Monotonicity principle
first, *58*, 63, 64, 81, 82, 120
second, *62*, 64, 84–86, 90, 113, 120, 144
Multiplicity of eigenvalue, *39*, 56, 78, 83

N

Natural boundary condition, 59
Nitsche, J. A., 115, 139
Nitsche, J. C., 139
Nonlinear problems
eigenvalues and, 6–18
linear theory and, 6–18
Non-self-adjoint problems, 149-50
Norm, *24*, 35, 55, 86, 88, 97, 107
of a bounded linear functional, 25
Sobolev, *28*, 46
Null space, *21*

O

Operator
completely continuous (compact), *53*
elliptic, 35
linear, *20*, 21
projection, 85, 108
self-adjoint, *37*
Order, *78*, 83
Orthogonal, *24*
complement, *82*, 87, 102
decomposition, 91
projection, 69, 80, 82, 91, 92, 96, 118–20, 131
Orthogonality of eigenvectors, 38–39
Orthonormal sequence, 51
Osborn, J. E., 149, 150

P

Parseval's equations, *52*, 53, 72, 73, 84
Payne, L. E., 115, 126, 143
Perturbation, 75–81
Pierce, J. G., 126, 132
Poincaré, H., 59
Poincaré's inequality, 45
Poincaré's principle, *55*, 56, 58
Pólya, G., 116, 128, 135, 143, 145, 148–49
Positive definite, *23*
Positive semidefinite, *23*
Pre-Hilbert space, *24*
completion of, 25–26, 29, 54
Projection, 69, 80, 82, 91, 92, 96, 118–20
Protter, M. H., 145
Pyramid function, 128, 135

Q

Quadratic functional (form), *22*
completely continuous, *50*
positive definite, *23*
positive semidefinite, *23*

R

Rabinowitz, P., 18, 35
Lord Rayleigh (J. W. Strutt), 59, 63
Rayleigh quotient, *45*, 57
eigenvalues of, *45*, 56
Rayleigh–Ritz method, 58, 59, 109, 149
approximating eigenvectors, 123–26
error bounds for eigenvalues in, 68, 70, 71, 73–74, 111, 116
finite difference equations from, 127–29
Reality of eigenvalues, 37–38
Regularity, 54, 73, 145
Resolvent
operator, *74*, 84
set, *74*
Reiss, E., 17
Ritz, W., 59

S

Sather, D., 17
Sattinger, D. H., 15
Scalar product
in Hilbert space, *25*
in pre-Hilbert space, *24*
Schiffer, M., 143, 145
Schroedinger equation, 6, 48
Schultz, M. H., 132
Schwarz's inequality, *23*
Self-adjoint, *37*
eigenvalue problem, 31–38
Separation of variables, 1–6, 10, 15, 16
Separation theorem, *63*, 64, 65, 84
Sesquilinear functional (form), *22*, 23, 31, 35, 53–55
Hermitian, *22*, 23, 31, 35, 54–55
Sobolev
norm, *28*, 73
space, *29*, 28–29, 46
Sorokin, V. S., 35
Sparse matrix, *128*, 134
Spectrum, *44*
continuous, *44*, 49, 50
discrete, *44*
essential, *44*
Spline functions, *131*, 132

Stability, 7, 8, 11, 13, 15, 17
Stable boundary condition, 59
Steiner symmetrization, *146*, 148
Stenger, W., 84
Strang, G., 132
Strong derivatives, *29*
Sturm–Liouville problems, 65
Sum of operators, 65
Swartz, B., 70, 126
Symmetric
  bilinear functional, *22*, 23
  integral operator, 115, 118
Symmetrization, 146–49
Synge, J. L., 128
Szegö, G., 148, 149

T

Temple, G., 115
Trefftz, E., 115, 116, 123, 126
Transformation, linear, *20*, 21
  adjoint, 118, 119
Triangle inequality, *24*
Truncation, 89–91, 94, 103, 105

U

Ukhovskii, M. R., 18
Unstable boundary condition, 59

V

Vainikko, G. M., 150
Vanish on the boundary of *D*, *46*
Varga, R. S., 126, 132
Vibrating beam, 13–15
Vibrating elastic solid, 15–16, 33–34, 60
Vibrating membrane, 32, 50, 59, 60, 61, 94, 105, 107, 118, 128, 134–39, 141–45, 148
Vibrating plate, 16–17, 59, 143
Vibrating string, 64, 71–73

W

Walker, A. G., 115
Washizu, K., 152
Weinberger, H. F., 65, 84, 91, 92, 113, 126, 134, 136, 139, 143, 145
Weinstein, A., 65, 74, 75, 77, 81, 83, 84
Weinstein–Aronszajn
  method, error estimates for, 91–98
  perturbation theorems, 74–81
Weinstein determinant, *75*, 78, 83, 86, 88, 91
Weinstein method, 81–84, 86–88, 89–91, 93, 98, 104–5, 121
Weinstein, D. H., 115
Wendroff, B., 70, 126, 136
Weston, J. D., 115
Weyl, H., 56, 63, 65
Wielandt, H., 115

Z

Zlamal, M., 132